Problems in Chemistry:

QUESTIONS AND SOLUTIONS

Ronald Ragsdale

University of Utah

Saunders Golden Sunburst Series

SAUNDERS COLLEGE PUBLISHING

Philadelphia New York Chicago
San Francisco Montreal Toronto
London Sydney Tokyo

Address orders to:
383 Madison Avenue
New York, NY 10017

Address editorial correspondence to:
West Washington Square
Philadelphia, PA 19105

PROBLEMS IN CHEMISTRY: QUESTIONS AND SOLUTIONS

ISBN # 0-03-011482-9

6789 018 987654321

CBS COLLEGE PUBLISHING
Saunders College Publishing
Holt, Rinehart and Winston
The Dryden Press

THIS BOOK IS DEDICATED TO THE MEMORY OF JEFFREY RONALD.

ACKNOWLEDGEMENT

The author greatly appreciates the patience and efforts of Donna Keyes who typed this manuscript.

CONTENTS

PREFACE

Most of the questions contained in this book have been asked on actual chemistry examinations. These questions have been compiled for students because most textbooks contain only exercises and problems. Yet most instructors of large enrollment courses give primarily multiple choice type of examinations. Students need practice on these types of questions.

Over 1000 questions have been compiled according to typical categories with answers. Besides the answers, most of the questions have complete solutions. A student can gain maximum benefit from this compilation if he/she **TRIES** to solve each question before looking at the answer or solution which is on the back of each page.

These questions provide good review and practice for exams such as the MCAT, DAT AND VCAT. Some questions are drill but many of them are teaching type of questions which provide the student with common incorrect answers. The student can discover the incorrect answers for himself/herself before taking an actual examination.

Many of the questions have listed with the answer the percent of the students who got the correct answer on an actual exam. By considering this information and comparing it to what he/she has just done, it is possible for each student to ascertain his/her preparation for an exam. The questions serve as a self-test for the student.

It has been my experience that many students will practice on these questions without the instructor having to assign them. These questions increase the time that a student spends on problem-solving.

Practice on these questions will greatly increase a student's understanding of chemistry. These questions provide a student with immediate feed-back so that they will readily do them on their own. Of course an instructor can always stimulate further activity by including a couple of the questions on an examination.

Students, it is important that you don't memorize answers. Discover the purpose of a question and develop an understanding of the principle(s) involved in the question. Many times an instructor will write new questions for an examination by looking at previous examination questions. The same concept is asked but just in a different way.

This book includes a special twelve page letter to the student on how to study chemistry. The student should read this letter **several times** during the year. It is important to develop good study habits and to develop new approaches to the study of chemistry. This book has been designed for you, the student.

THE UNIVERSITY OF UTAH

DEPARTMENT OF CHEMISTRY
HENRY EYRING BUILDING
SALT LAKE CITY, UTAH 84112

Dear Student:

I know that many of you are under a lot of pressure to do well in General Chemistry. Chemistry has been designated as a pre-engineering class; it is a prerequisite for medical school, dental school as well as a basic or required course for a number of disciplines. In most classes everyone can receive a high grade by "the old fashion way, by earning it." In fact, most instructors are willing to give all A's and B's or all A's if each student's performance is high enough. It is up to you.

It has been reported that 60% of the freshman who enter college never achieve their goals. Why? Because they have deficiencies in the basic learning skills -- reading, writing, listening, remembering, note taking, and in test taking. The good news is that these skills can be learned.

The pathway to success in college, in General Chemistry, and in any endeavor can be summarized by two words, **GOOD MANAGEMENT.** Most students who do poorly have usually had very bad management. Ben Hogan* has been reported to have said that golf is 80% management and 20% skill. For chemistry, it isn't any different as most students in this class have enough skill to be successful.

I have observed that good students usually have good management skills and poor students employ poor management. One area where many students exercise poor management is by not having a course load that is compatible with other activities or work. At the beginning of a new year or new term, too many students are too ambitious or optimistic about how much they can do. Few people can push themselves beyond 60 hours of rigorous, productive study and work for extended periods of time. For a student to be successful, he/she must plan that every credit hour will on the average require at least two hours outside of class plus one hour in class.

By considering the number of hours in outside activities, you can use Figure 1 to determine the number of credit hours that you can take. Remember your **study time** must be of high quality for it to count. Using the information from this figure, we can see that a student who carries 20 credit hours should not have a part-time job. A student who works 25 hours per week should not carry more than 11 - 12 credit hours.

*Ben Hogan was a 137 lb. golfer who predated Arnold Palmer and Jack Nicklaus. He won many of the major tournaments such as the Masters U.S. Open and the P.G.A.. Some of these tournaments were won after he was involved in a head-on collision with a bus. He was told that he would never walk again without a cane, but he came back by exercising good management with a tremendous determination. As a result, he was able to continue as one of the outstanding golfers in the world.

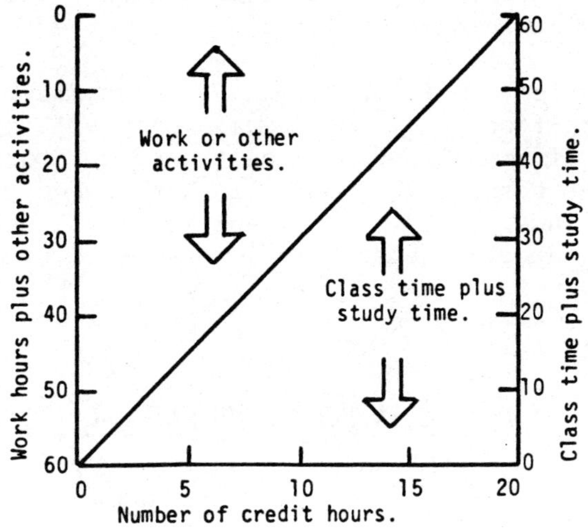

Fig. 1. A compatibility plot between a university class load and work load.

Many students (in particular, students who want to perform better in school than they have in the past) attempt to conquer chemistry by brute force. They try to overwhelm the course by greatly increasing their efforts. This approach is stimulated by past experience and is reflected in Figure 2. The rationale goes like the following:

> I can run twice as far in one hour as I can in 30 minutes.
> I can shovel twice as much snow in 30 minutes as I can in 15 minutes.
> My car goes twice as far on a full tank of gas as it can on a half a tank.

We have a linear relationship, and we equate this type of experience to learning. Unfortunately, many times in our studies, we do not have a linear relationship (we would like one), but we have a learning curve like that depicted in Figure 3. To be successful, we need to avoid getting on a flat portion of the curve, as we don't learn that much more with a large increase in time.

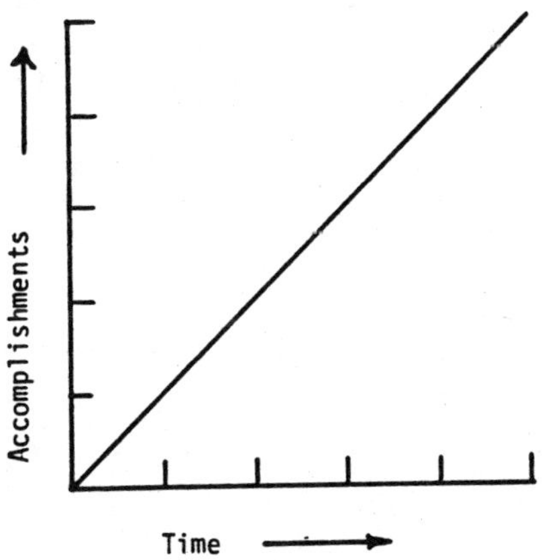

Fig. 2. The way we usually equate our accomplishments with time.

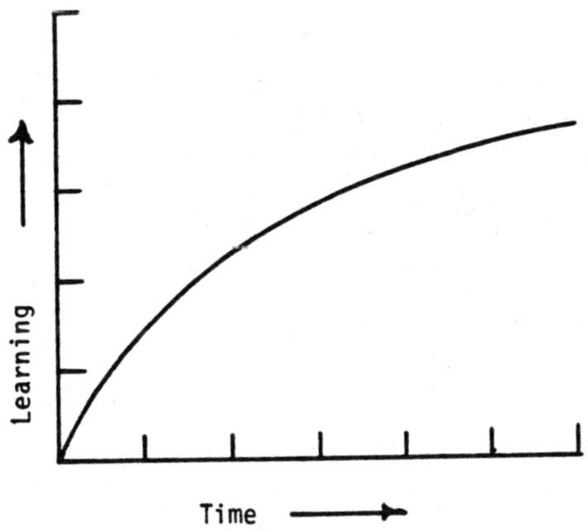

Fig. 3. A learning curve versus time.

One of the purposes of this letter is to stimulate you to develop new learning skills, better study habits, and better approaches to learning chemistry. This way you can avoid a flat portion of the learning curve. This will also help relieve some of your frustrations and anxieties. By good management, one should recognize when he/she is approaching diminishing returns on his/her efforts.

For many of you, this is your first term in a university. I consider the first term to be the most important one. If you do well, you will continue to do well. Likewise, I consider the first hour exam to be the most important exam that will be taken in any course. If you score high, you will then have a greater tendency to continue to do so on further exams. Why? You expect it of yourself, your friends expect it and your instructors expect it also. It is easier to study when you are encouraged and optimistic.

Now for some suggestions:

ALWAYS STUDY WITH A PENCIL IN YOUR HAND AND USE IT!!!

LECTURE:

Read assigned reading before coming to lecture.

Why? You will be sufficiently familiar with the material so you will more readily understand the lecture.

TAKING NOTES:

Try to capture the ideas and concept of a lecture.
Don't recopy your notes.

Why? One is not learning while meticulously copying material. It is time that is wasted.

Don't use a tape recorder.

Why? It takes too long to review the lecture. Take notes and review them.

Immediately after lecture (or as soon as possible) review your notes.

Why? In some research by Sonne, a group that reviewed immediately after lecture recalled one and a half times more than a group of students who did not review. Most forgetting takes place immediately after a lecture. Help eliminate this forgetting by reviewing. The successful learning process depends upon decelerating forgetting and accelerating learning. A typical forgetting curve is shown in Figure 4.

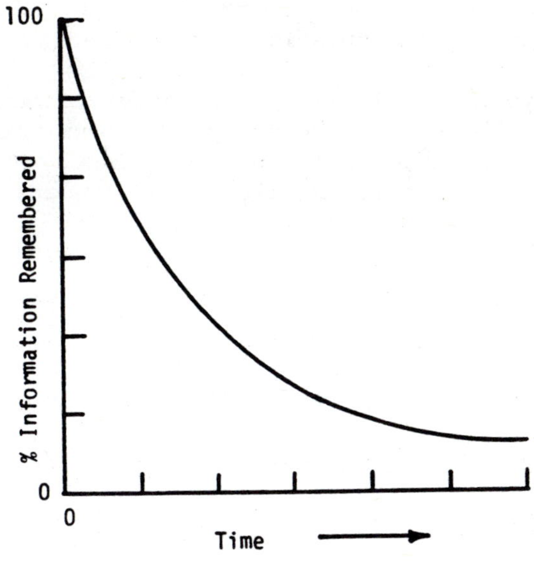

Fig. 4. Forgetting curve.

The shape of this curve can be changed by proper review.

Remembering what you have heard is usually more difficult than remembering what you have read. Therefore, take notes so that you will have something to read.

Don't do your remembering by memorizing.

Why? We need to be selective in what we memorize. William James said "The essence of genius is to know what to overlook." Herman Ebbinghaus found that it takes fifteen times as long to memorize twelve nonsence syllables as it does to learn a six syllable list. Time and effort is not a linear relationship, rather it is an exponential relationship. See Figure 3.

Memorization should only be done with material that is understood.

Why? Memorization should not be used as a means of escaping the effort to understand. Do not learn by memorizing what can be learned by reasoning. We can improve our memory by learning and understanding our chemistry initially. Then our review time will be meaningful.

The five R's of note-taking, the Cornell system:

1. Record. The meaningful ideas and concepts.
2. Reduce. After lecture, summarize the main ideas and concepts.
3. Recite. Say it out loud in your own words the main ideas of the lecture.
4. Reflect. Take a few minutes to ponder over the main ideas of the lecture.
5. Review. Once a week review the ideas of all the term's lectures.

ASKING QUESTIONS:

During lecture, ask questions of yourself.

Why? It helps to keep you mentally alert. It helps you to look for answers in the lecture. These same ideas apply to your reading. Ask questions while reading; this will help keep you alert. The best way to learn is to ask questions orally or silently. Have an inquiring mind.

Some questions that you should ask are the following:

What was the purpose of that demonstration?

What should I have learned from that demonstration?

What was the purpose of that example?

What was the purpose of that part of the lecture?

What should I have learned from that paragraph? From that section of the text?

How does this section fit in with the chapter?

How does the lecture fit in with the textbook?

Always have at least two questions that you would like to ask each time you go to your discussion class.

"President Corson of Cornell, formerly Dean of the College of Engineering, said that engineers and other students in science and mathematics must often "crack" the meaning of an idea or concept one sentence at a time. If comprehension does not occur even at this snail's pace, then ask the instructor for assistance. But before you do, ask yourself this question: "What is it that I don't understand?" Once the student clarifies what it is that he doesn't understand, he will have understood."[1]

Reference 2, Page 193

[1]Walter Pauk, How to Study in College, 2nd Ed. Houghton Mifflin Co., 1974, p. 146

[2]R. A. Carman and W. R. Adams, Jr., Study Skills, 2nd Ed., John Wiley and Sons, 198

MAKE A SCHEDULE:

Why? It is a means of gaining extra time. We can gain extra time in two ways: First, by being more efficient and second, by using small blocks of time that we usually waste.

Why? It helps get you started. It prevents you from avoiding your study.

Make each hour a productive time. Use daylight hours.

Why? Research shows that each hour used for study during the day is equal to one and a half hours at night.[3]

Study before each discussion class.

Why? The material will be fresh in your mind.

Study each subject in one hour block of time.

Why? It makes you more efficient. You don't become bored. Longer blocks of time lead to waste of time. Therefore, you will learn more chemistry by studying it every other hour.

Allow time for sleep.

Why? Sacrificing sleep in order to gain extra time for study can lead to a vicious circle.

Create the right atmosphere for cramming.

Why? If cramming for an exam is to be effective, the original studying and learning must take place each day. A good schedule facilitates this type of learning.

Decreases procrastination.

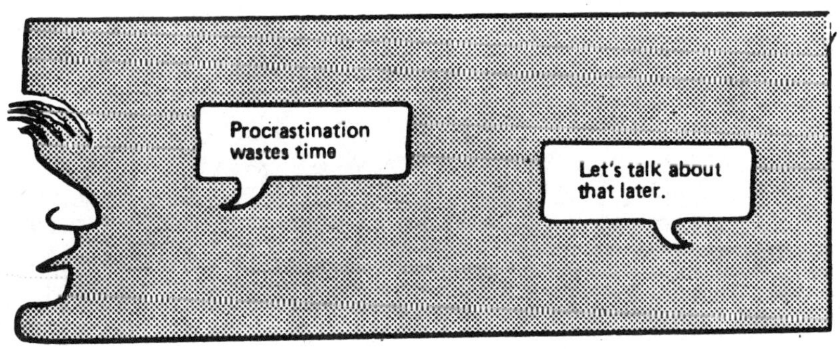

Reference 2, Page 69

[3]Reference 1, p. 22.

STUDYING:

Many students in this class are so desirous of being successful, that they too often use the brute force approach. See Figure 3, learning versus time.

Many times just before an examination, a student finds himself/herself on a flat portion of the learning curve. Thus, a big increase in time does not lead to much of an increase in learning. Consequently, it is important that a student is using his/her time effectively. Use good study skills so that you can change the shape of the curve in Figure 3.

Set realistic study goals for yourself.

Why? You can reward yourself for being successful. If you waste all afternoon and then set a goal of studying chemistry from 6:00 p.m. to midnight, you may not be realistic. The chances are that you won't be successful, you will become frustrated. Unfortunately, your one area of success may be in punishing yourself. That's not our goal. We want to learn chemistry.

Our studying must be efficient.

Why? We usually don't have enough time as it is.

Reference 2, Page 107

Never spend more than 30 minutes trying to solve a problem. After 15 minutes, any more time becomes a waste of time. Ask yourself these questions: What is it that I don't understand? Why can't I solve this problem? After 15 - 30 minutes, get HELP. See your teaching assistant. Interact with a fellow classmate. See a tutor.

Study at a desk, with your feet flat on the floor and with a pencil.

Why? It increases your learning efficiency. See again the figure on the Learning Curve.

Recite while studying.

Why? If you will recite while writing, you will be learning while using your eyes, ears, and muscles. If you can't say it, you don't know it. Say it in your own words. Explain a concept to someone else. "Students who silently read their assignments, no matter how diligently, often wonder why they remember so little after several days have passed. They do not realize that using the eyes alone as a means of absorbing information is an inefficient and ineffective method."[4]

"How recitation works. Recitation transfers material to the secondary memory. While you are reading the words in a sentence or paragraph, the primary memory (short-term memory) holds them in mind long enough for you to gain the sense or the sentence or the paragraph. However, the primary memory has a very limited capacity, and as you continue to read, you displace the words and ideas of the initial paragraphs with the words of subsequent paragraphs. This is one reason for not remembering everything in the first part of the chapter by the time we reach the end of the chapter when we read continually without pausing to recite.

It is only when we recite or contemplate the idea conveyed by a sentence or paragraph that the idea has a chance (not guaranteed) of moving on into the secondary memory (a long-term storage facility).

All verbal information goes first into the primary memory (short-term memory). When it is rehearsed (recited), part of it goes into our secondary (long-term) memory. The rest of it, usually the part we are least interested in, returns to the primary memory and is then forgotten.

Whether new information is "stored" or "dumped" depends, then on our reciting it out loud and on our interest in the information."[5]

Studying a textbook.

Use the SQ3R method of Francis P. Robinson:

> Survey: Glance over the different headings in a chapter. Get a perspective of what the chapter is about.
>
> Question: Turn each heading in a chapter to a question. This will help arouse your curiosity and thus increase your comprehension.
>
> Read: Read to answer your question.
>
> Recite: Recite the answer to your questions. Do something with the idea or phrase that you want to remember. Read it. Say it. See it. Speak it. Write it. The more you say it, the better you will remember it.

[4]Reference 1, p. 71.

[5]Reference 1, p. 70.

Review: Only by reviewing can you shift an idea into the long term memory bank.

Marking a textbook:

Don't mark a sentence until you have read a complete paragraph or a headed section.

Why? This keeps you from grabbing at every idea. Remember you cannot remember everything.

Don't overmark.

Be selective.

The following two sections on the next page are excerpts from Kotz and Purcell. The first section is an example of overmarking. The second section is a good example of how one should mark a page. Remember, study with a pencil.

Studying for an examination.

Learn with understanding. Don't memorize past questions. Understand what old questions were trying to measure. Take old questions and write some new questions utilizing the same concept.

Don't cram.

Why? It isn't effective unless previous study has taken place. Frantic study just before an exam is not as effective as regular study. Consider the study involving some rats. "Dr. Edward C. Tolman, after experimenting with rats over a long period of years, found that rats that learned to run a maze under the pressure of hunger took much longer to learn the maze than rats that learned under non-crisis conditions. Furthermore, the learning that did take place was of a narrow type; that is, after learning the "right" route, these rats panicked if one avenue were blocked off. They were not able to survey the field to notice alternative routes. On the other hand, when the rats were permitted to learn under non-crisis conditions, they later performed well in a crisis.

People react in similar ways. Under the pressure of an unannounced fire alarm, people in one experiment took two to three times as long to learn the escape route as they would under non-crisis conditions. It seems that the thinking mechanism is overstimulated and too jammed with unorganized thoughts, each of which counteracts the previous one, causing jerky, panicky reactions.

balanced chemical equation. The exponents, which can be zero or positive or negative whole numbers or fractions, must be determined by experiment. For example, experiment shows that the decomposition of H_2O_2 in the presence of iodide ion as catalyst has the rate law

$$\text{Initial rate} = k[H_2O_2]_0[I^-]_0$$

where the exponent on each concentration term is 1, even though the stoichiometric coefficient of H_2O_2 is 2 and I^- does not even appear in the balanced equation.

The symbol "k" is called the **rate constant** and is a proportionality constant relating rate and concentration. Unlike R, however, k must be evaluated for each reaction at each temperature.

Notice the word "initial" in the rate expression, and the 0 subscripts on the concentrations. The initial reaction rate is the instantaneous rate at the very beginning of the reaction,* as illustrated by Figure 13.4.

THE ORDER OF A REACTION

As you shall see in Section 13.7, a great deal of information is conveyed about reaction mechanisms by the "order" of a reaction. **The order** with respect to a particular reactant is the exponent of its concentration term in the rate law, and the **total reaction order** is the sum of the exponents on all the concentration terms.

An example of a **first order reaction** (a reaction with a total order of 1) is the change in the structure of a molecule. For example, azobenzene changes from one form (*cis*) to another (*trans*)

(Because the same ion, HCO_3^-, can be both a Brønsted acid and base, it is said to be **amphiprotic**.) In each reaction, the acceptance or donation of a proton by HCO_3^- produces a **conjugate** partner. Conjugate acids and bases are related to one another by the gain or loss of a proton. Every acid–base reaction involving H^+ transfer has two conjugate acid–base pairs (Section 15.2).

Acids and bases vary considerably in their ability to act as proton donors and acceptors (Section 15.2). HCl is a strong acid and is nearly 100% ionized in dilute aqueous solutions. In contrast, HF is only weakly ionized. This means that HCl has a great tendency to donate a proton *and* that its conjugate base, Cl^-, only weakly accepts a proton to return to HCl. Conversely, since HF is a weak acid, its conjugate base, F^-, is relatively strong, readily accepting a proton from water to re-form HF. These acids illustrate two corollaries of the Brønsted concept: (1) the stronger an acid or base, the weaker its conjugate base or acid, respectively. (2) In general, acid–base reactions proceed in the direction favoring the weaker acid–base pair (Table 15.2).

The concentration of hydronium ion in solution can be expressed in terms of pH (Section 15.3).

$$\text{pH} = -\log[H^+]$$

These same people, when given a fire drill in another building under non-crisis conditions learned the proper escape route quickly and well. After a few practice drills, these people could be counted on to act rationally under crisis conditions, not only finding the correct route promptly, but also finding an alternative route when necessary. This is the reason for fire drills in schools and especially on board ship.

This is a direct answer to the student who tries to learn under the pressure of cramming. During the crisis of the examination, on questions that require thinking beyond the facts or with different facts, the student will be panic-prone, and will often answer a question with wrong material, or will make many false, frantic starts, or will just "freeze". It seems that studying under intense pressure precludes later flexibility in thinking and acting."[6]

Taking an examination.

Always answer some easy questions first.

Why? This helps get you started and will help relieve test anxieties.

If you get stuck on a question, leave it.

Why? Later in the exam, an idea may come to you that will help you solve your problem. By leaving a question, you can help alleviate some of your frustration for not being able to solve it at that time.

You cannot usually expect to score 100%.

Why? The averages on many chemistry exams run from 50 to 70%.

Why do chemistry professors write such "challenging" exams? It helps spread the scores of the students. It makes it easier and fairer to assign grades, in particular to the A's and B's. There will be fewer students on the borderline. Most students will miss a number of questions on an examination, so don't let it disturb you while you are taking an exam if you can't solve all of the problems.

Reference 2, Page 248

[6]Reference 1, p. 175

Tutors.

Always get help before it is too late. Do not get a tutor who spends most of his time lecturing. Get one who forces you to answer questions. Remember, you learn the most when _you_ verbalize. By someone asking you questions, you discover what you don't know.

Studying with a friend.

This can be particularly advantageous because you have the opportunity to explain and to verbalize. Go to an empty classroom and work on a blackboard. Don't be a sponge, contribute to the study session. Teach someone else. **REMEMBER, IF YOU CAN'T SAY IT, YOU DON'T KNOW IT.**

Your teaching assistant.

See your T.A. Ask him/her questions. Remember, don't spend more than 15 minutes on a problem. Your T.A. has office hours and many times no one comes during their hours.

Your professor.

Your instructor is available during office hours and by appointment. See your instructor before you get into difficulty. They love to solve chemistry problems.

Good luck.

Sincerely,

Ronald O. Ragsdale
Professor

INTRODUCTION

Significant Figures

1.1 The number of significant figures in 30500 is

a. 1 b. 2 c. 3 d. 4 e. 5

1.2 The number of significant figures in 0.00650100 is

a. 3 b. 4 c. 6 d. 8 e. 9

1.3 The number of significant figures in 0.05070×10^{-2} is

a. 2 b. 3 c. 4 d. 5 e. 6

1.4 When combining the masses 0.0562 Kg, 124.213 g and 1635 mg, the total should be reported to _______ significant figures.

a. 3 b. 4 c. 5 d. 6 e. 7

1.5 Three samples were weighed using three different balances. (All are as accurate as the precision below indicates.) The masses are 0.924 kg, 75.9 g and 100. mg. The total mass should be reported as

a. 1000.000 g b. 1000.00 g c. 1000.0 g d. 1.000×10^3 g
e. 1.00×10^3 g

1.6 A physics student measured the electronic charge on one electron by using the Millikan oil droplet experiment and obtained the following three values:

4.635×10^{-10} esu 4.923×10^{-10} esu 4.768×10^{-10} esu

The best value for him to report is:

a. 4.7753×10^{-10} esu b. 4.775×10^{-10} esu
c. 4.78×10^{-10} esu d. 4.80×10^{-10} esu e. 4.8×10^{-10} esu

The answer to many of the questions is preceded by a number with a percent sign. This number represents the percent of students who were successful with this question on an actual examination. If you answer each question without looking at the answer, then the percentage number will help you to assess your level of preparation. The number will also indicate the difficulty level of a question.

70% 1.1 (c) The two trailing zeros indicate the order of magnitude of the number. It is best to use exponential notation 3.05×10^4 to indicate whether the zeros are significant figures.

95% 1.2 (c) The two zeros immediately to the right of the decimal point indicate the order of magnitude of the number 6.50100×10^{-3}.

90% 1.3 (c) A zero which is used to establish the order of magnitude is not significate. 5.070×10^{-4}

70% 1.4 (b)

$$\begin{array}{r} 56.\underline{2} \\ 124.21\underline{3} \\ 1.63\underline{5} \\ \hline 182.\underline{0}48 \end{array}$$

The total should be reported as 182.0 g, since the first mass was weighed to ±0.1 g, so the uncertainty is in the tenths.

65% 1.5 (d)

$$\begin{array}{r} 92\underline{4}.0 \\ 75.\underline{9} \\ 0.1\underline{00} \\ \hline 100\underline{0}.000 \end{array}$$

The uncertainty is in the units place.

40% 1.6 (e) The uncertainty in the student's measurements is occurring in the tenths position. It would probably be best for him to report the value as $(4.8 \pm 0.2) \times 10^{-10}$.

1.7 If the volume and mass measurements on a sample of copper at 25°C are 32.465 g and 3.62 mL, the value for the density should have how many significant figures?

a. one b. two c. three d. four e. five

1.8 The number of significant figures that should be reported for the mass of a sample which is obtained by taking the difference between the mass of a sample plus beaker (79.451 g) and the mass of the beaker (78.83 g) is

a. 1 b. 2 c. 3 d. 4 e. 5

1.9 The best answer to report for $\frac{2.481 \times 12.74}{2.69} + 0.27$ is

a. 12.0202 b. 12.020 c. 12.02 d. 12.0 e. 12

1.10 The term that refers to the reproducibility of a laboratory measurement is

a. precision b. repeatability c. accuracy d. exactness

e. the number of significant figures

Conversions

1.11 The heat of combustion of ethanol is -29.8 kJ g^{-1}. Expressed in J kg^{-1}, the heat of combustion is

a. -2.98×10^{1} J kg^{-1} b. -2.98×10^{3} J kg^{-1}

c. -2.98×10^{4} J kg^{-1} d. -2.98×10^{7} J kg^{-1}

e. -2.98×10^{10} J kg^{-1}

1.12 The heat of combustion of glucose, $C_6H_{12}O_6$, is -15600 J g^{-1}. Expressed in kJ kg^{-1}, the heat of combustion is

a. -1.56×10^{1} kJ kg^{-1} b. -1.56×10^{4} kJ kg^{-1}

c. -1.56×10^{6} kJ kg^{-1} d. -1.56×10^{7} kJ kg^{-1}

e. -1.56×10^{9} kJ kg^{-1}

95% 1.7 (c) For multiplication and division, the product or quotient should not possess any more significant figures than the least precisely known factor in the calculation.

55% 1.8 (b)

$$\begin{array}{r} 79.45\underline{1} \\ -78.8\underline{3} \\ \hline 0.6\underline{2}1 \end{array}$$

When adding or subtracting, the number of decimal places in the answer should be equal to the number of decimal places in the number with the fewest places.

60% 1.9 (d)

$$\underbrace{\frac{2.481 \times 12.74}{2.69}} + 0.27$$

Because of this operation, there can only be three significant figures in the final answer. See the explanation to question 1.7.

85% 1.10 (a) Accuracy is how close the answer is to the true value. Precision indicates how close the experimental measurements are to each other.

50% 1.11 (d) $-29.8 \frac{kJ}{g} \times 1000 \frac{g}{kg} \times 1000 \frac{J}{kJ}$

45% 1.12 (b) $-15600 \frac{J}{g} \times \frac{1\ kJ}{1000\ J} \times \frac{1000\ g}{1\ kg}$

1.13 How many 100. mg terramycin capsules can be made from 10.0 kg of terramycin?

a. 1000 b. 10,000 c. 100,000 d. 1,000,000

e. 10,000,000

1.14 The average velocity of carbon dioxide molecules at 0°C is 3.62×10^4 cm/sec. Which of the indicated calculations below would convert this value to the velocity in miles per hour?

a. $\frac{1 \text{ sec}}{3.62 \times 10^4 \text{ cm}} \times \frac{2.54 \text{ cm}}{1 \text{ in}} \times \frac{12 \text{ in}}{1 \text{ ft}} \times \frac{5280 \text{ ft}}{1 \text{ mi}} \times \frac{1 \text{ hr}}{3600 \text{ sec}}$

b. $\frac{1 \text{ sec}}{3.62 \times 10^4 \text{ cm}} \times \frac{2.54}{\text{cm}} \times \frac{1 \text{ ft}}{12 \text{ in}} \times \frac{1 \text{ mi}}{5280 \text{ ft}} \times \frac{3600 \text{ sec}}{1 \text{ hr}}$

c. $3.62 \times 10^4 \frac{\text{cm}}{\text{sec}} \times \frac{2.54 \text{ cm}}{1 \text{ in}} \times \frac{12 \text{ in}}{1 \text{ ft}} \times \frac{5280 \text{ ft}}{1 \text{ mi}} \times \frac{3600 \text{ sec}}{1 \text{ hr}}$

d. $3.62 \times 10^4 \frac{\text{cm}}{\text{sec}} \times \frac{1 \text{ in}}{2.54 \text{ cm}} \times \frac{12 \text{ in}}{1 \text{ ft}} \times \frac{5280 \text{ ft}}{1 \text{ mi}} \times \frac{1 \text{ hr}}{3600 \text{ sec}}$

e. $3.62 \times 10^4 \frac{\text{cm}}{\text{sec}} \times \frac{1 \text{ in}}{2.54 \text{ cm}} \times \frac{1 \text{ ft}}{12 \text{ in}} \times \frac{1 \text{ mi}}{5280 \text{ ft}} \times \frac{3600 \text{ sec}}{1 \text{ hr}}$

1.15 How many scruples are there in 100. lbs? Some equivalents which may be helpful are given below:

1.00 scruple = 20.0 grains, 1.00 g = 15.4 grains,

1.00 grain = 0.0648 g, 1.00 lb = 453.6 g, 1.00 Kg = 2.205 lb.

a. $\frac{1.00 \text{ scruple}}{20.0 \text{ grains}} \times \frac{1.00 \text{ grain}}{0.0648 \text{ g}} \times \frac{1.00 \text{ Kg}}{2.205 \text{ lb}} \times \frac{1.00 \text{ lb}}{453.6 \text{ g}} \times \frac{1000 \text{ g}}{1 \text{ Kg}} \times 100. \text{ lbs.}$

b. $\frac{20.0 \text{ grains}}{1.00 \text{ scruple}} \times \frac{1.00 \text{ g}}{15.4 \text{ grains}} \times \frac{1.00 \text{ grain}}{0.0648 \text{ g}} \times \frac{453.6 \text{ g}}{1.00 \text{ lb}} \times 100. \text{ lbs.}$

c. $\frac{1.00 \text{ scruple}}{20.0 \text{ grains}} \times \frac{15.4 \text{ grains}}{1.00 \text{ g}} \times \frac{0.0648 \text{ g}}{1.00 \text{ grain}} \times \frac{453.6 \text{ g}}{1.00 \text{ lb}} \times 100. \text{ lbs.}$

d. $\frac{1.00 \text{ scruple}}{20.0 \text{ grains}} \times \frac{15.4 \text{ grains}}{1.00 \text{ g}} \times \frac{453.6 \text{ g}}{1.00 \text{ lb}} \times 100. \text{ lbs.}$

e. $\frac{1.00 \text{ scruple}}{20.0 \text{ grains}} \times \frac{1.00 \text{ grain}}{0.0648 \text{ g}} \times \frac{1.00 \text{ Kg}}{2.205 \text{ lb}} \times 100. \text{ lbs.}$

55% 1.13 (c) $$\frac{10.0\ \text{kg} \times \frac{1000\ \text{g}}{1\ \text{kg}} \times \frac{1000\ \text{mg}}{1\ \text{g}}}{100.\ \text{mg/capsule}}$$

90% 1.14 (e) $$3.62 \times 10^{4}\ \frac{\text{cm}}{\text{sec}} \times \frac{1\ \text{in}}{2.54\ \text{cm}} \times \frac{1\ \text{ft}}{12\ \text{in}} \times \frac{1\ \text{mi}}{5280\ \text{ft}} \times \frac{3600\ \text{sec}}{1\ \text{hr}}$$

By cancelling the units, one can see that we are left with mi/hr, the desired unit.

70% 1.15 (d) The final desired unit is scruples.

$$\frac{1.00\ \text{scruple}}{20.0\ \text{grains}} \times \frac{15.4\ \text{grains}}{1.00\ \text{g}} \times \frac{453.6\ \text{g}}{1.00\ \text{lb}} \times 100.\ \text{lbs}$$

Please note that in this problem we did not use all the conversion factors which were given in the stem of the problem.

1.16 The distance from Salt Lake City to Los Angeles is 700 miles by car. Which of the following is a correct set-up for that distance in kilometers?

1 in = 2.54 cm
1 mile = 5280 ft
1 foot = 12 inches
1 m = 100 cm
1 km = 1000 m

a. $700 \times \frac{12}{1} \times \frac{2.54}{1} \times \frac{1}{1000}$ km

b. $700 \times \frac{5280}{1} \times \frac{12}{1} \times \frac{1}{100} \times \frac{2.54}{1} \times \frac{1}{1000}$ km

c. $700 \times \frac{5280}{1} \times \frac{1}{12} \times \frac{2.54}{1} \times \frac{100}{1} \times \frac{1}{1000}$ km

d. $700 \times \frac{5280}{1} \times \frac{12}{1} \times \frac{1}{2.54} \times \frac{1}{100} \times \frac{1}{1000}$ km

e. $700 \times \frac{12}{1} \times \frac{1}{2.54} \times \frac{1760}{1} \times \frac{1}{1000}$ km

1.17 How many firkins are there in 825 in^3? Some conversion factors which may be helpful are:

1 barrel = 4 firkins
1 gallon = 231 in^3
1 gallon = 3.78 liters
1 barrel = 42.0 gallons

a. 4.12×10^{-4} firkins b. 0.340 firkins c. 1.38 firkins
d. 72.7 firkins e. 2430 firkins

50% 1.16 (b) This problem was made more difficult because the units were not given with the conversion factors. Always show the units for each factor.

$$700 \text{ mi} \times \frac{5280 \text{ ft}}{1 \text{ mi}} \times \frac{12 \text{ in}}{1 \text{ ft}} \times \frac{1 \text{ m}}{100 \text{ cm}} \times \frac{2.54 \text{ cm}}{1 \text{ in}} \times \frac{1 \text{ km}}{1000 \text{ m}}$$

65% 1.17 (b) $$\text{firkins} = 825 \text{ in}^3 \times \frac{4 \text{ firkins}}{1 \text{ barrel}} \times \frac{1 \text{ barrel}}{42 \text{ gal}} \times \frac{1 \text{ gal}}{231 \text{ in}^3}$$

One of the purposes of this problem is to show that you can solve problems even when you are not acquainted with the units in question, if you will use unit analysis.

1.18 A piece of iron with a volume of 1.452 gal weighs 95.31 lbs. Calculate the density of iron in scruple/drachm. Some equivalents which may be helpful are: 1 L = 0.264 gal, 1 kg = 2.2046 lb., 1 scruple = 1.296 g, 1 metric ton = 1000 kg, 1 mL = 0.2816 drachm, 1 m = 1000 mm.

a. 5.00×10^{-3} b. 21.5 c. 272 d. 309 e. 531

Physical and Chemical Properties

1.19 All of the following are intensive properties EXCEPT

a. density b. temperature c. heat d. melting point
e. freezing point

1.20 In addition to volume, which property of matter must be known to calculate mass?

a. temperature b. molecular weight c. density d. pressure
e. specific heat

1.21 Which of the following is not a characteristic property of a compound?

a. molecular formula b. melting point c. mass d. density
e. heat of vaporization

1.22 A crystal clear colorless liquid is heated in an open beaker and stirred vigorously. Visible boiling, as indicated by bubble formation, begins at 101°C and rises up to 106°C while boiling continues. The heating is stopped at 106°C. Based on the information given, we can say that

a. the material in the beaker was a pure compound.
b. the material in the beaker was a pure element.
c. the material in the beaker was a pure substance.
d. the material in the beaker was a homogeneous mixture.
e. the material in the beaker was a heterogeneous mixture.

35% 1.18 (b) This problem was made more difficult because there was the necessity of changing both the mass unit and volume unit since d = m/V.

$$\frac{95.31 \text{ lbs} \times \frac{1.000 \text{ kg}}{2.205 \text{ lb}} \times \frac{1000 \text{ g}}{1 \text{ kg}} \times \frac{1.000 \text{ scruple}}{1.296 \text{ g}}}{1.452 \text{ gal} \times \frac{1.00 \text{ L}}{0.264 \text{ gal}} \times \frac{1000 \text{ mL}}{1 \text{ L}} \times \frac{0.2816 \text{ drachm}}{1.000 \text{ mL}}}$$

85% 1.19 (c) An extensive property is related the amount of substance whereas an intensive property is not related to the amount of substance.

90% 1.20 (c) $d = \frac{m}{V}$; $m = d \times V$

70% 1.21 (c) The molecular formula, melting point, density, and heat of vaporization are intensive properties.

50% 1.22 (d) The boiling point of a pure substance is constant. Since the boiling point is changing with time, we must have a mixture. The more volatile component is boiling off first. The solution is homogeneous because the original solution was clear.

1.23 All the following are characteristic properties of phosphorus. Which one is a chemical property?

a. Both red phosphorus and white phosphorus exist in solid allotropic forms.

b. The red form melts at about 600°C and the white form melts at 44°C.

c. The white form is soluble in liquid carbon disulfide, but is insoluble in water.

d. When exposed to air, white phosphorus will burn spontaneously, but red phosphorus will not.

e. The red form of phosphorus is insoluble in both water and carbon disulfide.

1.24 All of the following are examples of physical properties EXCEPT

a. water freezing b. iodine evaporating c. gasoline burning

d. sea water boiling e. mercury expanding

1.25 Which of the following is an example of a chemical change?

a. Sublimation of dry ice. b. Alcohol boiling.

c. Melting of sodium chloride. d. Burning of hydrogen.

e. Dissolving of sucrose.

1.26 A sample of carbon dioxide which undergoes transformation from solid to liquid to gas would undergo

a. a change in density. b. a change in mass.

c. a change in composition. d. no change in physical properties.

e. no change in energy.

1.27 Bromine is a liquid at room temperature and it has a density of 3.12 g/mL at 25°C. What volume is occupied by 50.0 g of bromine?

a. 0.0624 mL b. 8.01 mL c. 16.0 mL d. 50.0 mL

e. 156 mL

$3.12 = \frac{50}{V}$

85% 1.23 (d) A chemical property involves a chemical change (a reaction).

90% 1.24 (c) A chemical property involves a change in properties. A rearrangement in the atoms to form new molecules.

90% 1.25 (d) Sublimation, boiling, melting and dissolving are physical changes.

1.26 (a)

1.27 (c) $V = \frac{m}{d} = \frac{50.0\ g}{3.12\ g/mL}$

1.28 Calculate the weight of iron which occupies the same volume as 85 g of aluminum. Densities: Fe = 7.87 g/ml, Al = 2.70 g/ml

a. 29.2 g   b. 31.5 g   c. 247.8 g   d. 248 g   e. 2.5×10^2 g

Atomic Symbols

1.29 Which one of the following lists gives the correct symbols for the elements bromine, boron, chlorine and silicon in that order?

a. Be, Bo, Co, S   b. Br, Bo, Cr, S   c. Br, B, Cl, Si

d. Br, B, Cl, S   e. Be, B, Cr, Si

1.30 Which one of the following lists gives the symbols for the elements phosphorus, potassium, silver, chlorine, and sulfur in that order?

a. K, Ag, Po, Cl, S   b. P, Po, Ag, Cl, S   c. Ph, K, Ag, S, Cl

d. P, K, Ag, Cl, S   e. Ph, Po, Ag, Cl, S

1.31 Which of the following elements has the greatest atomic weight?

a. magnesium   b. manganese   c. cadmium   d. calcium

e. chromium

Temperature Scales

1.32 Aluminum melts at 660.°C. What is its melting point in °F?

a. 349 °F   b. 1160 °F   c. 1220 °F   d. 1340 °F   e. 1430 °F

1.33 Liquid propane boils at -42°C. What is its boiling point on the kelvin scale?

a. 231 K   b. 356 K   c. 273 K   d. 315 K   e. 345 K

1.34 Liquid ethane boils at -89°C. What is its boiling point on the kelvin scale?

a. 164 K   b. 184 K   c. 298 K   d. 304 K   e. 362 K

35% 1.28 (e) $V_{Al} = \frac{85 \text{ g Al}}{2.70 \text{ g/mL}} = 31.5 \text{ mL}$

$m_{Fe} = 7.87 \text{ g/mL} \times 31.5 \text{ mL} = 2.5 \times 10^2 \text{ g}$

The final answer has 2 significant figures because the mass of Al was only given to two significant figures.

100% 1.29 (c)

95% 1.30 (d) P is phosphorus and K is potassium (K comes from the Latin word kalium).

90% 1.31 (c) Students frequently interchange Mg and Mn. A student should know the symbols of the common elements.

75% 1.32 (c) $°F = 1.8\ °C + 32\ °F$

$°F = 1.8 \times 660 + 32$

95% 1.33 (a) $K = °C + 273\ K$

$K = -42 + 273$

95% 1.34 (b) $K = -89 + 273$

1.35 Liquid oxygen boils at 90. K. What is its boiling point on the Fahrenheit temperature scale?

a. -361 °F b. -329 °F c. -297 °F d. -196 °F e. -183 °F

60% 1.35 (c) $°C = K - 273$

$$°C = 90 - 273 = -183$$

$$°F = 1.8\ °C + 32$$

$$°F = 1.8\ (-183) + 32$$

2

STOICHIOMETRY

Laws of Chemical Combination

2.1 Dalton's atomic theory consisted of all the following postulates EXCEPT

a. atoms of different elements having different properties.

b. elements are composed of indivisible particles called atoms.

c. with gases the volumes consumed and produced are small whole number ratios.

d. atoms combine in fixed ratios of whole numbers.

e. in chemical changes atoms are not destroyed, created, or changed.

2.2 From the thermal decomposition of a pure solid one obtained a solid and a gas, each of which is a pure substance. From this information, we can conclude with certainty that

a. the original solid is not an element.

b. at least one of the products is an element.

c. both products are elements.

d. the solid is a compound and the gas is an element.

e. the solid is an element and the gas is a compound.

2.3 Solid ammonium chloride decomposes upon heating to ammonia and hydrogen chloride. The mass of hydrogen chloride and ammonia equals the original mass of the ammonium chloride. This illustrates the Law of

1. Constant Composition
2. Multiple Proportions
3. Conservation of Mass

a. 1 only b. 2 only c. 3 only d. 1 and 2 only e. 1, 2 and 3

2.4 Which of the following sets could be used to illustrate the law of multiple proportions?

a. O, O_2, O_3 b. ${}^{1}_{1}H$, ${}^{2}_{1}H$, ${}^{3}_{1}H$ c. MgO, CaO, BeO

d. NaF, MgF_2, AlF_3 e. ClF, ClF_3, ClF_5

70% 2.1 (c) Gay-Lussac discovered the law of combining volumes of gases in 1808. Avogadro provided the explanation, equal volumes of gases at the same temperature and pressure contain the same number of molecules.

2.2 (a) The original solid must have contained at least two different elements as two different products were formed.

2.3 (c) $NH_4Cl(s) \rightarrow NH_3(g) + HCl(g)$

$53.5\ (g) \rightarrow 17.0\ g \quad 36.5\ (g)$

75% 2.4 (e) The masses of one element that can combine chemically with a fixed mass of another element are in a small whole number ratio. Some other examples are CO and CO_2 and SO_2 and SO_3.

2.5 Analyses of two different compounds showed that each compound contained only sulfur and oxygen.

Compound 1 - 16.0 g of sulfur combined with 16.0 g of oxygen.

Compound 2 - 32.0 g of sulfur combined with 48.0 g of oxygen.

These data illustrate

1. The Law of Conservation of Matter.
2. The Law of Definite Composition.
3. The Law of Multiple Proportions.

a. 1 only b. 2 only c. 3 only d. 1 and 2 only e. 1, 2 and 3

2.6 All of the following <u>may</u> change during a chemical reaction EXCEPT

a. the total number of atoms in the system.
b. the temperature of the system.
c. the density of the system.
d. the total number of molecules in the system.
e. the total volume of the system.

Percent Composition

2.7 What is the percent nitrogen in ammonium carbonate, $(NH_4)_2CO_3$?

a. 14.53% b. 27.83% c. 29.16% d. 33.34% e. None of these.

2.8 What is the percentage of weight of oxygen in silver sulfate, Ag_2SO_4? Atomic weights: Ag - 107.9, S - 32.0, O - 16.0.

a. 10.3 b. 20.5 c. 31.3 d. 43.5 e. 79.5

2.9 A crystal of the mineral pharmacosiderite, $Fe_3(AsO_4)_2(OH)_3 \cdot 5H_2O$ (FW = 586.5) contains ________% oxygen by mass:

a. 43.6% b. 38.2% c. 30.0% d. 27.3% e. 13.6%

2.10 A crystal of the compound $UO_2(C_2H_3O_2)_2 \cdot NH_4C_2H_3O_2 \cdot 6\ H_2O$ (FW = 573) contains ________% oxygen by weight:

a. 5.58 b. 6.98 c. 16.8 d. 22.3 e. 39.1

2.5 (c) We don't know if the Law of Definite Composition is applicable because only one sample of each compound was analyzed. There is a small whole-number ratio between the masses of oxygen (SO_2 and SO_3).

88% 2.6 (a) Conservation of mass.

The reaction may be exothermic or endothermic. Reactions involve the decomposition of molecules and the making of molecles, so there could be a different number of molecules.

2.7 (c) $\% \text{ N} = \frac{2 \times 14.01}{96.08} \times 100 = 29.16\%$

2.8 (b) $\% \text{ O} = \frac{4 \times 16.0}{2 \times 107.9 + 32.0 + 4\,(16.0)} \times 100$

2.9 (a) $\% \text{ O} = \frac{16 \times 16.0}{586.5} \times 100$

88% 2.10 (e) $\% \text{ O} = \frac{14 \times 16.0}{573} \times 100$

The Mole and Avogadro's Number

2.11 How many atoms are in 12 molecules of glucose, $C_6H_{12}O_6$?

a. 24 b. 288 c. 2160 d. 7.22×10^{24} e. 1.73×10^{26}

2.12 In 0.500 mole of acetic acid (ethanoic acid), CH_3COOH, there are:

a. 6.02×10^{23} molecules b. 1.2×10^{24} molecules

c. 1.8×10^{24} atoms d. 2.4×10^{24} atoms e. 4.8×10^{24} atoms

2.13 Which of the following does not describe 44.0 g of propane, C_3H_8?

a. One mole of propane.

b. The amount of propane that contains 8.0 g of hydrogen.

c. The amount of propane that contains $8 \times 6.02 \times 10^{23}$ hydrogen atoms.

d. The amount of propane that contains 36.0 g of carbon.

e. $44.0 \times 6.02 \times 10^{23}$ molecules of propane.

2.14 Calculate the number of atoms in 4.0×10^{-5} g of aluminum.

a. 8.9×10^{17} b. 4.6×10^{19} c. 6.5×10^{20} d. 3.8×10^{23}

e. 1.1×10^{29}

2.15 Which of the following samples contains the smallest number of atoms?

a. 1 g H_2 b. 1 g O_2 c. 1 g O_3 d. 1 g S_8 e. 1 g Cl_2

2.16 Which of the following samples contains the largest number of atoms?

a. 1 g Li b. 1 g F_2 c. 1 g C d. 1 g P_4 e. 1 g N_2

2.17 How many moles of fluorine gas (F_2 molecules) are present in 55.0 g of fluorine? Atomic weight of fluorine = 19.0.

a. 1.45 b. 1.71 c. 2.42 d. 2.75 e. 2.89

2.11 (b) 12 molecules $C_6H_{12}O_6 \times \frac{24 \text{ atoms}}{1 \text{ molecule}} \; C_6H_{12}O_6$

2.12 (d) $0.500 \text{ mole} \times 6.02 \times 10^{23} \frac{\text{molecules}}{\text{mole}} \times \frac{8 \text{ atoms}}{\text{molecule}}$

2.13 (e) 44.0 g of propane is equivalent to 1 mole of propane which is equivalent to Avogadro number of molecules.

2.14 (a) $4.0 \times 10^{-5} \text{ g Al} \times \frac{1 \text{ mole Al}}{27.0 \text{ g Al}} \times 6.02 \times 10^{23} \frac{\text{atoms}}{\text{mole}}$

44% 2.15 (e) All this question is asking is which element has the largest atomic mass. It should not be necessary to do any calculations.

64% 2.16 (a) This question is asking which element is the lightest.

76% 2.17 (a) $\# \text{ mol } F_2 = 55.0 \text{ g } F_2 \times \frac{1 \text{ mol } F_2}{38.0 \text{ g } F_2}$

2.18 Which of the following samples contains 2.0×10^{23} atoms?

a. 9.0 g O_2 b. 13.0 g K c. 15.0 g P_4 d. 19.0 g F_2

e. 12.0 g Mg

2.19 What is the total number of atoms (not molecules) present in 25.0 <u>mg</u> of citral (an organic compound that is used in the synthesis of vitamin A)? It has the molecular formula $C_{10}H_{16}O$.

a. 9.89×10^{19} b. 9.89×10^{20} c. 2.32×10^{21} d. 2.67×10^{21}

e. 9.89×10^{22}

2.20 Sodium cyclamate, $C_6H_{11}NHSO_3Na$, was used at one time as an artificial sweetner. If $C_6H_{11}NHSO_3Na$ has a molecular mass of 201.2 g/mol, how many moles of sodium cyclamate are contained in a 25.6 g sample?

a. 0.127 mol b. 0.193 mol c. 0.245 mol d. 7.90 mol

e. 5180 mol

2.21 A sample of fructose, a fruit sugar, $C_6H_{12}O_6$, that contains 24 g of carbon also contains ___g of oxygen.

a. 24 g b. 32 g c. 48 g d. 64 g e. 96 g

2.22 Which one of the following has the greatest mass?

a. An HF molecule b. 20.0 g of HF c. 10.0 mol of H_2

d. 1.0 mol of F_2 e. 2.0 mol of H_2O

2.23 How many moles of <u>nitrogen</u> gas (N_2 molecules) are present in 48.0 grams of nitrogen? Atomic weight N = 14.0.

a. 0.58 moles b. 0.86 moles c. 1.71 moles d. 2.00 moles

e. 3.42 moles

80% 2.18 (b) $\# \text{ atoms} = 13.0 \text{ g K} \times \frac{1 \text{ mol K}}{39.1 \text{ g K}} \times \frac{6.0 \times 10^{23} \text{ atoms}}{1 \text{ mol}}$

44% 2.19 (d) $\# \text{ atoms} = 25.0 \text{ mg} \times \frac{1 \text{ g}}{1000 \text{ mg}} \times \frac{1 \text{ mol}}{152 \text{ g}} \times \frac{6.0 \times 10^{23} \text{ molecules}}{1 \text{ mol}} \times \frac{27 \text{ atoms}}{1 \text{ molecule}}$

95% 2.20 (a) $\# \text{ mol} = 25.6 \text{ g} \times \frac{1 \text{ mol}}{201.2 \text{ g}}$

88% 2.21 (b) # mol C = # mol O (6 C and 6 O)

$\# \text{ mol C} = 24 \text{ g C} \times \frac{1 \text{ mol C}}{12.0 \text{ g C}} = 2.0$

$\text{wt. O} = 2.0 \text{ mol O} \times \frac{16.0 \text{ g O}}{1 \text{ mol O}}$

2.22 (d) One may have to do two or three calculations on a problem like this.

$1 \text{ mol } F_2 \times \frac{38.0 \text{ g}}{1 \text{ mol}} = 38.0 \text{ g}$

$2.0 \text{ mol } H_2O \times \frac{18.0 \text{ g}}{1 \text{ mol}} = 36.0 \text{ g}$

68% 2.23 (c) $48.0 \text{ g } N_2 \times \frac{1 \text{ mole } N_2}{28.0 \text{ g } N_2}$

28% of the students selected answer (e), so they used the atomic mass of N and not the molecular mass of N_2.

2.24 How many atoms are present in 435 mg of KPF_6 (MW = 184.1)?

a. 1.42×10^{21} b. 9.96×10^{21} c. 1.14×10^{22} d. 1.42×10^{24}

e. 1.14×10^{25}

2.25 The weight in grams of a <u>single</u> molecule of CO_2 (C = 12.0; O = 16.00) is:

a. 28 g b. 44 g c. 6.0×10^{-23} g d. 7.31×10^{-23} g

e. 7.3×10^{23} g

2.26 What is the mass of 0.216 mol of $C_8H_9O_4$?

a. 36.5 g b. 56.4 g c. 86.9 g d. 169 g e. 782 g

Balancing Equations and the Mole Concept

2.27 Balance the following expression:

$$\_\_ NH_3 + \_\_ O_2 \rightarrow \_\_ NO_2 + \_\_ H_2O$$

The <u>balanced</u> equation shows that 1.00 mole of NH_3 requires ___ mole(s) of O_2:

a. 0.57 b. 1.25 c. 1.33 d. 1.75 e. 2.67

2.28 The products of the combustion of acetaldehyde with oxygen are shown in the equation below:

$$\_\_ CH_3CHO + \_\_ O_2 \rightarrow \_\_ CO_2 + \_\_ H_2O$$

When properly balanced, the equation indicates that ___ mole(s) of O_2 are required for each mole of CH_3CHO.

a. 1 b. 2 c. 2.5 d. 3 e. 5

68% 2.24 (c) $435 \text{ mg} \times \frac{1 \text{ g}}{1000 \text{ mg}} \times \frac{1 \text{ mol}}{184.1 \text{ g}} \times \frac{6.02 \times 10^{23} \text{ molecules}}{1 \text{ mol}} \times \frac{8 \text{ atoms}}{1 \text{ molecule}}$

64% 2.25 (d) Mass one molecule $= 44.0 \frac{\text{g}}{\text{mole}} \times \frac{1 \text{ mole}}{6.02 \times 10^{23} \text{ molecules}}$

$= 7.31 \text{ g/molecule}$

32% of the students selected the mass of one mole of CO_2; 44.0 g/mole.

2.26 (a) $0.216 \text{ mol} \times 169.2 \frac{\text{g}}{\text{mol}} = 36.5 \text{ g}$

92% 2.27 $4\ NH_3 + 7\ O_2 \rightarrow 4\ NO_2 + 6\ H_2O$

$\#\ \text{mol}\ O_2 = \frac{7\ O_2}{4\ NH_3} \times 1 \text{ mol } NH_3$

2.28 (c) $2\ CH_3CHO + 5\ O_2 \rightarrow 4\ CO_2 + 4\ H_2O$

$1\ CH_3CHO + 5/2\ O_2 \rightarrow 2\ CO_2 + 2\ H_2O$

Don't forget the oxygen atom in acetaldehyde otherwise you would obtain (d) as an answer.

2.29 Balance the following equation with the SMALLEST WHOLE NUMBER COEFFICIENTS possible. Select the number that is the sum of the coefficients in the balanced equation:

$$\_\_KClO_3 \rightarrow \_\_KCl + \_\_O_2$$

a. 5 b. 6 c. 7 d. 8 e. 9

2.30 When the expression

$$C_5H_6NS(\ell) + O_2(g) \rightarrow CO_2(g) + H_2O + NO_2(g) + SO_2(g)$$

is balanced, the sum of all the smallest whole number coefficients is

a. 28 b. 29 c. 33 d. 36 e. 39

2.31 The complete combustion of ethane yields carbon dioxide and water. When the equation below is balanced, the ratio of the coefficients of CO_2 to H_2O is

$$\_\_\ C_2H_6(g) + \_\_\ O_2(g) \rightarrow \_\_\ CO_2(g) + \_\_\ H_2O(\ell)$$

a, 1:1 b. 2:3 c. 3:2 d. 1:3 e. 3:1

2.32 The complete combustion of butanone yields carbon dioxide and water. Balance the equation with the smallest whole number coefficients possible. Select the number that is the sum of the coefficients in the balanced equation.

$$\_\_\ CH_3CH_2COCH_3(\ell) + \_\_\ O_2(g) \rightarrow \_\_\ CO_2(g) + \_\_\ H_2O(\ell)$$

a. 15 b. 17 c. 27 d. 29 e. 30

In questions 2.33 and 2.34, balance the equations with the smallest whole number coefficients possible. In each case, choose the number that is the sum of the coefficients in the balanced equation. <u>Don't forget coefficients of "one"</u>.

2.33 $\_\_Ca_3(PO_4)_2 + \_\_C \rightarrow \_\_Ca_3P_2 + \_\_CO$

a. 14 b. 16 c. 18 d. 20 e. 24

2.34 $\_\_Nb_2O_5 + \_\_CCl_4 \rightarrow \_\_NbCl_5 + \_\_COCl_2$

a. 12 b. 13 c. 16 d. 18 e. 21

2.29 (c) $2\ KClO_3 \rightarrow 2\ KCl + 3\ O_2$

68% 2.30 (e) $2\ C_5H_6NS + 17\ O_2 \rightarrow 10\ CO_2 + 6\ H_2O + 2\ NO_2 + 2\ SO_2$

80% 2.31 (b) $2\ C_2H_6 + 7\ O_2 \rightarrow 4\ CO_2 + 6\ H_2O$

$4\ CO_2 : 6\ H_2O$

56% 2.32 (d) $2\ CH_3CH_2COCH_3 + 11\ O_2 \rightarrow 8\ CO_2 + 8\ H_2O$

88% 2.33 (c) $1\ Ca_3(PO_4)_2 + 8\ C \rightarrow 1\ Ca_3P_2 + 8\ CO$

92% 2.34 (b) $1\ Nb_2O_5 + 5\ CCl_4 \rightarrow 2\ NbCl_5 + 5\ COCl_2$

2.35 Balance the following expression:

$$\_\_\ CH_3CH_2COOH + \_\_\ O_3 \rightarrow \_\_\ CO_2 + \_\_\ H_2O$$

The balanced equation shows that 3.00 moles of propanoic acid requires ___ moles of ozone, O_3, for complete combustion.

a. 2 b. 3 c. 6 d. 7 e. 9

Questions 2.36 and 2.37 refer to the following reaction:

$$C_3H_8 + 5\ O_2 \rightarrow 3\ CO_2 + 4\ H_2O$$

2.36 How many moles of water will be produced from the complete combustion of 4.4 g of C_3H_8? Atomic weights: Carbon = 12.00, H = 1.00.

a. 0.10 moles b. 0.25 moles c. 0.40 moles d. 1.00 moles

e. None of the above; the answer is _________.

2.37 How many grams of oxygen are required to burn the 4.4 grams of C_3H_8? (Atomic weight oxygen = 16.0.)

(a) 8.0 b. 12 c. 16 d. 32 e. 64

2.38 BaO_2, barium peroxide, decomposes when heated to give BaO and O_2. Write a balanced equation for this reaction. If 0.500 mole of BaO_2 is decomposed, the number of moles of O_2 formed is:

a. 8.00 b. 4.00 c. 1.00 d. 0.500 e. 0.250

Determination of Chemical Formulas

2.39 In a particular apatite ore in which $Ca(OH)_2$ is associated with calcium phosphate, $Ca_3(PO_4)_2$, an analysis shows that calcium and phosphorus are present in a mol ratio of 5:3. Which of the following formulae agrees with this information?

a. $Ca(OH)_2 \cdot Ca_3(PO_4)_2$ b. $Ca(OH)_2 \cdot 2Ca_3(PO_4)_2$ c. $Ca(OH)_2 \cdot 3Ca_3(PO_4)_2$

d. $(Ca(OH)_2)_2 \cdot Ca_3(PO_4)_2$ e. $(Ca(OH)_2)_3 \cdot Ca_3(PO_4)_2$

2.35 (d) $3\ CH_3CH_2COOH\ +\ 7\ O_3\ \rightarrow\ 9\ CO_2\ +\ 9\ H_2O$

68% 2.36 (c) $\#\text{ moles } H_2O = \frac{4\ H_2O}{1\ C_3H_8} \times 4.4\text{ g } C_3H_8 \times \frac{1\text{ mole } C_3H_8}{44.0\text{ g } C_3H_8}$

56% 2.37 (c) $\#\text{ moles } O_2 = \frac{5O_2}{1\ C_3H_8} \times 4.4\text{ g } C_3H_8 \times \frac{1\text{ mole } C_3H_8}{44.0\text{ g } C_3H_8} = 0.50\text{ mole}$

$\#\text{ g } O_2 = 0.50\text{ mole } O_2 \times \frac{32.0\text{ g } O_2}{1\text{ mole } O_2} = 16\text{ g } O_2$

28% of the students selected 8.0 g, so they took the atomic mass of oxygen instead of the molecular mass.

80% 2.38 (e) $2\ BaO_2(s) \xrightarrow{\Delta} 2\ BaO(s)\ +\ O_2(g)$

$\#\text{ moles } O_2 = \frac{1\ O_2}{2\ BaO_2} \times 0.500\text{ mole } BaO_2$

76% 2.39 (c) 10 Ca : 2 P

2.40 Of the following, the only empirical formula is

a. N_2F_2 b. N_2F_4 c. H_2C_2 d. H_2N_2 e. HNF_2

2.41 Of the following, the only empirical formula is

a. $C_6H_{12}O_6$ b. $C_{12}H_{22}O_{11}$ c. C_2H_4 d. C_6H_6 e. $C_2H_4O_2$

2.42 A compound consists of the following elements by weight percent:

Carbon - 40.0%

Oxygen - 53.3%

Hydrogen - 6.7%

The ratio of carbon to oxygen to hydrogen in the empirical formula is

a. 1:2:1 b. 1:1:1 c. 1:1:2 d. 2:1:2 e. 2:1:1

2.43 A compound contains 54.53% carbon, 9.15% hydrogen, and 36.32% oxygen. Atomic weights: C = 12.00, H = 1.01, O = 16.00. What is the number of hydrogens in the simplest formula of the compound (empirical formula)?

a. 2 b. 4 c. 5 d. 6 e. 9

2.44 Analysis of a compound yields the following percent composition by weight: 17.55% Na, 39.70% Cr, 42.75% O. What is the simplest formula of this compound? Atomic weights: Na = 23.0, Cr = 52.0, O = 16.0.

a. Na_2CrO_4 b. $Na_2Cr_2O_7$ c. $NaCrO_2$ d. $NaCr_2O_3$ e. $Na_2Cr_2O_3$

2.45 Analysis of a hydrocarbon showed that it contained 14.4% hydrogen and 85.6% carbon by weight. What is its simplest formula?

a. CH b. CH_2 c. CH_3 d. C_2H_3 e. C_2H_5

2.46 In another experiment, it was found that the mass of one mole of the compound in problem 2.45 was 138 ± 4 g. What is the molecular formula for the compound?

a. C_9H_{18} b. $C_{10}H_{15}$ c. $C_{10}H_{20}$ d. $C_{10}H_{25}$ e. $C_{11}H_{11}$

84% 2.40 (e)

Molecular Formula	Empirical Formula
N_2F_2	NF
N_2F_4	NF_2
H_2C_2	HC
H_2N_2	HN

76% 2.41 (b) See solution for question 2.40.

2.42 (c) $\# \text{ mol C} = 40.0 \text{ g C} \times \frac{1 \text{ mol C}}{12.0 \text{ g C}} = 3.33$ 1

$\# \text{ mol O} = 53.3 \text{ g O} \times \frac{1 \text{ mol O}}{16.0 \text{ g O}} = 3.33$ 1

$\# \text{ mol H} = 6.7 \text{ g H} \times \frac{1 \text{ mol H}}{1.01 \text{ g H}} = 6.63$ 2

80% 2.43 (b) $\# \text{ mol C} = 54.53 \text{ g C} \times \frac{1 \text{ mol C}}{12.0 \text{ g C}} = 4.54$

$\# \text{ mol H} = 9.15 \text{ g H} \times \frac{1 \text{ mol H}}{1.01 \text{ g H}} = 9.06$

$\# \text{ mol O} = 36.32 \text{ g O} \times \frac{1 \text{ mol O}}{16.0 \text{ g O}} = 2.27$

$C \frac{4.54}{2.27} \quad H \frac{9.06}{2.27} \quad O \frac{2.27}{2.27}$; C_2H_4O

72% 2.44 (b)

92% 2.45 (b) $\# \text{ mol C} = 85.6 \text{ g C} \times \frac{1 \text{ mol C}}{12.01 \text{ g C}} = 7.13$

$\# \text{ mol H} = 14.4 \text{ g H} \times \frac{1 \text{ mol H}}{1.01 \text{ g H}} = 14.26$

$C \frac{7.13}{7.13} \quad H \frac{14.26}{7.13}$; CH_2

84% 2.46 (c) Formula mass of CH_2 = 14; # empirical units = $\frac{138}{14} = 9.86 = 10$

$\therefore C_{10}H_{20}$

The actual molecular mass is 140.3 g/mol.

2.47 A molecular compound contains 92.3% carbon and 7.7% hydrogen. If the molecular weight of the compound is 78.1, what is its molecular formula?

a. CH b. C_2H_2 c. C_5H_6 d. C_6H_6 e. C_6H_7

2.48 An organic compound which has the empirical formula CHO has a relative molecular mass of 232. Its molecular formula is:

a. CHO b. $C_2H_2O_2$ c. $C_3H_3O_3$ d. $C_4H_4O_4$ e. $C_8H_8O_8$

2.49 The molecular mass of a compound with the empirical formula, C_3H_4O, was measured by a student in the laboratory. She reported the value of 170 ± 10 g/mol. The molecular formula of this compound is

a. C_3H_4O b. $C_6H_8O_2$ c. $C_6H_{12}O_3$ d. $C_9H_{12}O_3$ e. $C_9H_{16}O_3$

2.50 A 27.0 g sample of an unknown carbon-hydrogen compound was burned in excess oxygen to form 88.0 g of CO_2 and 27.0 g H_2O. What is a possible molecular formula of the hydrocarbon?

a. CH_4 b. C_2H_6 c. C_4H_6 d. C_4H_8 e. C_4H_{10}

2.51 A certain compound has a molar mass of 56 g mol^{-1}. All the following are possible simplest formulas for this compound EXCEPT

a. CH_2 b. CO c. C_3H_4O d. CH_2O e. CH_2N

2.52 A compound contains by weight, 41.4% carbon, 3.47% H, and 55.1% oxygen. A 0.050-mole sample of this compound weighs 8.71 g. The molecular formula of the compound is:

a. CHO b. C_3H_3O c. $C_3H_3O_3$ d. $C_4H_4O_4$ e. $C_6H_6O_6$

2.53 When $CaSO_4 \cdot y\ H_2O$ is heated, all of the water is driven off. If 34.0 g of $CaSO_4$ (FW = 136) is formed from 43.0 g of $CaSO_4 \cdot y\ H_2O$, what is the value of y?

a. 1 b. 2 c. 3 d. 4 e. 5

2.47 (d) The empirical formula is CH ; $\frac{78}{13} = 6$

$\therefore C_6H_6$

OR $78.1 \times 0.923 = 72.1$ g C ; $72.1 \text{ g C} \times \frac{1 \text{ mol C}}{12.01 \text{ g C}} = 6$ C, etc.

96% 2.48 (e) Empirical mass = 12 + 1 + 16 = 29

empirical units = 232/29 = 8

92% 2.49 (d) Empirical mass = 3 x 12 + 4 x 1 + 1 x 16 = 56

empirical units = 170/56 = 3

Molecular mass = 9 x 12 + 12 x 1 + 3 x 16 = 168

48% 2.50 (c) # mol C = # mol CO_2 = $88.0 \text{ g } CO_2 \times \frac{1 \text{ mol } CO_2}{44.0 \text{ g } CO_2} = 2.0$

g C = $2.0 \text{ mol C} \times \frac{12.0 \text{ g C}}{1 \text{ mol C}} = 24.0$ g

g H = 27.0 - 24.0 = 3.0 g H

$\therefore$ empirical formula = C_2H_3

64% 2.51 (d) All the distractors are multiples of 56 except CH_2O which has a mass of 30.

2.52 (e) Empirical formula is CHO ; Molecular mass = $\frac{8.71 \text{ g}}{0.050 \text{ mol}} = 174.2$ g

174.2 / 29 = 6

2.53 (b) # mol $CaSO_4$ = $34 \text{ g } CaSO_4 \times \frac{1 \text{ mol}}{136 \text{ g}} = 0.25$

mol H_2O = $(43.0 - 34.0) \frac{1 \text{ mol}}{18.0 \text{ g}} = 0.50$

$(CaSO_4) \frac{0.25}{0.25} \quad (H_2O) \frac{0.50}{0.25}$

2.54 Sixteen grams of sulfur (atomic weight = 32) combine with a metal (atomic weight = 40.0) to give a product which weighs 26.0 g. What is the empirical formula of the sulfide formed. M = metal.

a. M_2S b. MS c. MS_2 d. M_3S_2 e. M_2S_3

2.55 An unknown element Z combines with oxygen to form the compound ZO_2. If 48.0 g of element Z combines exactly with 32.0 g of oxygen, what is the atomic weight of Z in g?

a. 24.0 b. 32.0 c. 48.0 d. 64.0 e. 96.0

Calculations and Chemical Equations (Including Limiting Reagent Problems)

2.56 Calculate the mass of hydrogen formed when 25 g of aluminum reacts with excess hydrochloric acid.

$$2\ Al + 6\ HCl \rightarrow 2\ AlCl_3 + 3\ H_2$$

a. 0.41 g b. 0.92 g c. 1.2 g d. 1.8 g e. 2.8 g

2.57 Carbon dioxide, CO_2, and ammonia, NH_3, combine together to form urea, CH_4ON_2, plus water. Write a balanced equation and calculate the mass of ammonia in grams that would be needed to make two moles of urea.

a. 25 to 35 g b. 35 to 45 g c. 50 to 60 g d. 60 to 70 g
e. 75 to 85 g

Questions 2.58 and 2.59 refer to the following reaction:

$$C_5H_{12} + 8\ O_2 \rightarrow 5\ CO_2 + 6\ H_2O$$

2.58 How many moles of water will be produced from the complete combustion of 14.4 g of C_5H_{12}? Atomic weights: Carbon = 12.01, H = 1.01.

a. 0.20 b. 0.60 c. 1.2 d. 2.4 e. 6.0

76% 2.54 (c)

$$\begin{array}{lrl} M_xS_y \text{ (product)} & = & 26.0 \text{ g} \\ \underline{-S_y \text{(reactant)}} & = & \underline{16.0 \text{ g}} \\ M \text{ (reactant)} & & 10.0 \text{ g} \end{array}$$

$$\# \text{ mol S} = 16.0 \text{ g S} \times \frac{1 \text{ mol S}}{32.0 \text{ g S}} = 0.50$$

$$\# \text{ mol M} = 10.0 \text{ g M} \times \frac{1 \text{ mol M}}{40.0 \text{ g M}} = 0.25$$

$$M \frac{0.25}{0.25} \quad S \frac{0.50}{0.25} \quad ; \quad MS_2$$

2.55 (c) One mole of Z combines with one mole of O_2. Therefore, the atomic mass of Z is 48.0 g/mol.

$$\# \text{ mol Z} = \frac{1 \text{ Z}}{2 \text{ O}} \times \# \text{ mol O} = \frac{1}{2} \times 32.0 \text{ g} \times \frac{1 \text{ mol O}}{16.0 \text{ g O}}$$

$$\text{At. Wt.} = \frac{48.0 \text{ g}}{1 \text{ mol}}$$

2.56 (e)

$$\# \text{ mol } H_2 = \frac{3 \text{ } H_2}{2 \text{ Al}} \times \# \text{ mol Al} = \frac{3}{2} \times 25 \text{ g Al} \times \frac{1 \text{ mol Al}}{27.0 \text{ g Al}} = 1.39$$

$$\text{mass } H_2 = 1.39 \text{ mol } H_2 \times 2.02 \text{ g } H_2/\text{mol } H_2 = 2.8 \text{ g } H_2$$

2.57 (d) $CO_2 + 2\ NH_3 \rightarrow CH_4ON_2 + H_2O$

$$\# \text{ } NH_3 = \frac{2 \text{ } NH_3}{1 \text{ } CH_4ON_2} \times 2 \text{ mol } CH_4ON_2 = 4.0 \text{ mol}$$

$$\text{Wt. } NH_3 = 4.0 \text{ mol } NH_3 \times \frac{17.0 \text{ g } NH_3}{\text{mol } NH_3} = 68.0 \text{ g}$$

68% 2.58 (c)

$$\# \text{ mol } H_2O = \frac{6 \text{ } H_2O}{1 \text{ } C_5H_{12}} \times \# \text{ mol } C_5H_{12}$$

$$= \frac{6}{1} \times 14.4 \text{ g } C_5H_{12} \times \frac{1 \text{ mol}}{72.0 \text{ g}}$$

2.59 How many grams of oxygen are required to burn the 14.4 g of C_5H_{12}?

a. 1.60 b. 12.8 c. 25.6 d. 51.2 e. 76.8

2.60 For the reaction

$$2\ MnO_2 + 4\ KOH + O_2 + Cl_2 \rightarrow 2\ KMnO_4 + 2\ KCl + 2\ H_2O$$

there is 100. g of each reactant available. Which reagent is the limiting reagent? (Formula weights are MnO_2 = 86.9; KOH = 56.1; O_2 = 32.0; Cl_2 = 70.9.)

a. MnO_2 b. KOH c. O_2 d. Cl_2

2.61 The overall reaction for the production of sodium dichromate can be represented by the equation

$$\underset{224}{\overset{360\text{ g}}{4FeCr_2O_4}} + \underset{106}{\overset{290\text{ g}}{8Na_2CO_3}} + \underset{32.0}{\overset{90\text{ g}}{7O_2}} + \underset{98.1}{\overset{180\text{ g}}{4H_2SO_4}} \rightarrow 4Na_2Cr_2O_7 + 4H_2O + 2Fe_2O_3 + 8CO_2 + 4Na_2SO_4$$

The formula weight for each reactant is shown below the reactant and the mass of each reactant for this problem is given above the reactant. Which reactant is the limiting reagent?

a. $FeCr_2O_4$ b. Na_2CO_3 c. O_2 d. H_2SO_4

2.62 What number of moles of O_2 is needed to produce 11.0 g of P_4O_6 from phosphorus? (Molecular mass P_4O_6 = 219.9)

a. 0.0500 mol b. 0.100 mol c. 0.150 mol d. 0.300 mol
e. 0.450 mol

2.63 How many g of nitric acid can be prepared from the reaction of 92.0 g of nitrogen dioxide with 36.0 g of water?

$$3\ NO_2 + H_2O \rightarrow 2\ HNO_3 + NO$$

a. 64 b. 76 c. 84 d. 116 e. 252

56% 2.59 (d) $\# \text{ mol } O_2 = \frac{8\ O_2}{1\ C_5H_{12}} \times 14.4 \text{ g } C_5H_{12} \times \frac{1 \text{ mol } C_5H_{12}}{72.0 \text{ g } C_5H_{12}}$

$$\text{wt. } O_2 = 1.6 \text{ mol } O_2 \times \frac{32.0 \text{ g } O_2}{1 \text{ mol } O_2}$$

76% 2.60 (b) Determine # mol of each reactant. Divide this number by the coefficient that appears before the reactant in the balanced equation. The smallest number is the limiting reagent.

$$\# \text{ mol } MnO_2 = 100 \text{ g} \times \frac{1 \text{ mol } MnO_2}{86.9 \text{ g } MnO_2} = 1.15 \; ; \quad \frac{1.15}{2} = 0.58$$

$$\# \text{ mol KOH} = 100 \text{ g} \times \frac{1 \text{ mol KOH}}{56.1 \text{ g KOH}} = 1.78 \; ; \quad \frac{1.78}{4} = 0.45$$

$$\# \text{ mol } O_2 = 100 \text{ g} \times \frac{1 \text{ mol } O_2}{32.0 \text{ g } O_2} = 3.13 \; ; \quad \frac{3.13}{1} = 3.13$$

$$\# \text{ mol } Cl_2 = 100 \text{ g} \times \frac{1 \text{ mol } Cl_2}{70.9 \text{ g } Cl_2} = 1.41 \; ; \quad \frac{1.41}{1} = 1.41$$

2.61 (b) A method for solving this type of question is given in question 2.60 above.

76% 2.62 (c) $\# \text{ mol } O_2 = \frac{3\ O_2}{1\ P_4O_6} \times \# \text{ mol } P_4O_6 = \frac{3}{1} \times 11.0 \text{ g } P_4O_6 \times \frac{1 \text{ mol } P_4O_6}{220 \text{ g } P_4O_6}$

2.63 (c) Which reactant NO_2 or H_2O is the limiting reactant?

$$\# \text{ mol } NO_2 = \frac{3\ NO_2}{1\ H_2O} \times \# \text{ mol } H_2O$$

$$\# \text{ mol } NO_2 = \frac{3}{1} \times 36.0 \text{ g } H_2O \times \frac{1 \text{ mol } H_2O}{18.0 \text{ g } H_2O} = 6 \text{ mol}$$

$$\text{wt. of 6 mol } NO_2 = 6 \text{ mol } NO_2 \times \frac{46.0 \text{ g } NO_2}{1 \text{ mol } NO_2} = 276 \text{ g}$$

$\therefore$ NO_2 is the limiting reagent.

$$\# \text{ mol } HNO_3 = \frac{2\ HNO_3}{3\ NO_2} \times \# \text{ mol } NO_2$$

$$= 2/3 \times 92.0 \text{ g } NO_2 \times \frac{1 \text{ mol } NO_2}{46.0 \text{ g } NO_2} = 1.33$$

$$\text{wt. } HNO_3 = 1.33 \text{ mol } HNO_3 \times \frac{63.0 \text{ g } HNO_3}{1 \text{ mol } HNO_3}$$

2.64 When 8.00 grams of hydrogen gas reacts with 32.0 grams of oxygen gas in an explosion, the final gas mixture will contain: (Atomic weights: H = 1.0, O = 16.0)

a. H_2, H_2O, and O_2 b. H_2, H_2O c. O_2, H_2O d. H_2, O_2

e. Pure H_2O

2.65 Nitric oxide, NO, is made from the oxidation of NH_3 and is represented by the equation:

$$4NH_3 + 5O_2 \rightarrow 4NO + 6H_2O$$

An 8.5 g sample of NH_3 gives 12.0 g of NO. The PERCENT YIELD of NO is: (Mol. wt. NH_3 = 17.0 ; Mol. wt. NO = 30.0)

a. 10.% b. 40.% c. 60.% d. 80.% e. 100.%

2.66 Sulfur trioxide, SO_3, is made from oxidation of SO_2 and is represented by the equation:

$$2\,SO_2 + O_2 \rightarrow 2\,SO_3$$

A 16.0 g sample of SO_2 gives 18.0 g of SO_3. THE PERCENT YIELD OF SO_3 is

a. 60% b. 75% c. 80% d. 90% e. 100%

70% 2.64 (b) $2\ H_2(g) + O_2(g) \rightarrow 2\ H_2O(g)$

$$\#\ \text{mol} = 8.00\ \text{g}\ H_2 \times \frac{1\ \text{mol}\ H_2}{2.0\ \text{g}\ H_2} = 4.0$$

$$\#\ \text{mol} = 32.0\ O_2 \times \frac{1\ \text{mol}\ O_2}{32.0\ \text{g}\ O_2} = 1.0$$

∴ O_2 is the limiting reagent and it will be completely consumed, leaving excess hydrogen and the product.

95% 2.65 (d) $\#\ \text{mol}\ NH_3 = \#\ \text{mol NO} = 8.5\ \text{g}\ NH_3 \times \frac{1\ \text{mol}\ NH_3}{17.0\ \text{g}\ NH_3}$

$\#$ mol NO = 0.50 moles

$$\text{Theoretical yield NO} = 0.50\ \text{moles} \times \frac{30.0\ \text{g NO}}{1\ \text{mol NO}}$$

Theoretical yield NO = 15.0 g

$$\%\ \text{yield} = \frac{12.0\ \text{g}}{15.0\ \text{g}} \times 100$$

80% 2.66 (d) $\text{Theoretical yield in mol of}\ SO_3 = \frac{2\ SO_3}{2\ SO_2} \times \#\ \text{mol}\ SO_2$

$$= 16.0\ \text{g}\ SO_2 \times \frac{1\ \text{mol}\ SO_2}{64.0\ \text{g}\ SO_2} = 0.25$$

$$\text{wt}\ SO_3 = 0.25\ \text{mol}\ SO_3 \times \frac{80.0\ \text{g}\ SO_3}{1\ \text{mol}\ SO_3} = 20.0\ \text{g}$$

$$\%\ \text{yield} = \frac{18.0\ \text{g}}{20.0\ \text{g}} \times 100$$

3

ATOMIC STRUCTURE

Atoms, Isotopes, and Ions

3.1 An element is

a. a collection of atoms with identical number of neutrons.

b. a collection of atoms with identical nuclear masses.

c. a collection of ions and atoms with identical number of electrons.

d. a collection of atoms with identical number of protons.

e. a collection of atoms all of which are identical in every way.

3.2 The number of protons in the nucleus determines the

a. atomic mass. b. atomic number. c. mass number.

d. number of neutrons. e. number of isotopes.

3.3 The number of neutrons in the nucleus of an atom of ${}^{19}_{9}F$ is

a. 9 b. 10 c. 18 d. 19 e. 28

3.4 The following species F^-, Ne, Na^+ and Mg^{2+} all have the same number of

a. protons b. nucleons c. isotopes d. neutrons

e. electrons

3.5 An atom that has the same number of neutrons as ${}^{52}Cr$ is

a. ${}^{50}Cr$ b. ${}^{52}Mn$ c. ${}^{56}Fe$ d. ${}^{54}Fe$ e. ${}^{59}Co$

85% 3.1 (d) All the atoms of a given element have the same atomic number. Therefore, they have the same number of protons.

96% 3.2 (b)

96% 3.3 (b) # neutrons = mass number - atomic number

= 19-9

72% 3.4 (e) F^-, Ne, Na^+ and Mg^{2+} are all isoelectronic as each has 10 electrons. (# nucleons = # neutrons + # protons)

56% 3.5 (d) $^{52}_{24}Cr$ has 28 neutrons (52 - 24)

$^{54}_{26}Fe$ has 28 neutrons (54 - 26)

3.6 Which one of the following statements about the isotopes of a given element is correct?

a. The atoms have the same number of protons but differing number of neutrons

b. The atoms have the same number of neutrons but differing numbers of electrons.

c. The atoms have the same number of electrons but differing numbers of protons.

d. The atoms have the same number of neutrons but differing numbers of protons.

e. The atoms have the same number of protons but differing numbers of electrons.

3.7 How many electrons does the ion, ${}^{56}_{26}Fe^{3+}$, have?

a. 23 b. 26 c. 29 d. 53 e. 56

3.8 Which of the following does not have a noble gas configuration?

a. P^{3-} b. Al^{3+} c. F^{-} d. K^{+} e. Zn^{2+}

Some Significant Contributions to the Development of Atomic Structure

Questions 3.9 - 3.11 consist of five letter headings which are followed by the three questions. For each question, select the one heading which is most closely related to it.

a. Thomson b. Rutherford c. Moseley d. Heisenberg

e. deBroglie

3.9 Characterized alpha, beta and gamma radiation.

3.10 An accurate determination of both the momentum and position of an electron is not possible.

3.11 Electrons can be considered to have both particle character and wave character.

68% 3.6 (a)

84% 3.7 (a) 26-3 = 23

3.8 (e) P^{3-} and K^+ have the Ar configuration. Al^{3+} and F^- have the Ne configuration. Zn^{2+} has the configuration $[Ar]3d^{10}$.

3.9 (b) Rutherford received the Nobel Prize for this work.

3.10 (d) Heisenberg received the Nobel Prize for this work.

3.11 (e) deBroglie received the Nobel Prize for this work.

For questions 3.12 - 3.14 there are five lettered headings followed by the numbered phrases. For each numbered phrase, select the one heading which is most closely related to it.

a. Millikan b. Bohr c. Einstein d. Rutherford

e. Heisenberg

3.12 Energy is radiated only when an electron falls from a higher to a lower energy level.

3.13 Discovered the nuclear atom.

3.14 Determined the charge of an electron by means of the oil droplet experiment.

3.15 X-rays are generated when a beam of cathode rays strikes a solid target. X-rays of different frequency result if the target is changed. Moseley's study of this phenomenon established that

a. the atomic number is a fundamental property of the atom (charge on the nucleus).

b. the mass of the atom is concentrated in its nucleus.

c. the value of e/m is 1.76×10^{8} coulombs/g.

d. the charge of the electron is 1.6×10^{-19} coulombs/g.

e. isotopes of elements exist.

3.16 Experiments were carried out in which a beam of CATHODE RAYS were first bent by a magnetic field then bent back by an electrostatic field until the beam hit the screen EXACTLY where it had been hitting the screen before the fields were applied. This experiment permitted the direct measurement of

a. the charge on the nucleus of an atom.

b. the charge on the electron, e.

c. the ratio of charge to mass of an electron, e/m.

d. the mass of an electron, m.

e. the mass of the atom.

3.12 (b) Bohr received the Nobel Prize for this work.

3.13 (d) Rutherford discovered the nuclear atom after he had received the Nobel prize.

3.14 (a) Millikan received the Nobel Prize for this work.

80% 3.15 (a)

92% 3.16 (c) Cathode rays are beams of electrons.

3.17 In Millikan's OIL DROPLET experiment, the charge on the oil droplet was always found to be a multiple of a given charge. Suppose the experiment was repeated in the laboratory and gave the following values for charge on <u>different</u> oil droplets.

Experiment	Charge Measured on Droplet
1	-2.40×10^{-19} charge units
2	-3.20×10^{-19} charge units
3	-5.60×10^{-19} charge units
4	-6.40×10^{-19} charge units
5	-7.20×10^{-19} charge units

On the basis of these data, we can conclude that the droplet with the smallest charge contained how many electrons?

a. 1 b. 3 c. 4 d. 5 e. 7

3.18 Rutherford's scattering experiment demonstrated

a. the existence of alpha and beta particles.
b. that most of the mass of an atom is in its nucleus.
c. the existence of electrons.
d. the existence of neutrons.
e. the charge mass ratio of an electron is constant.

Light, Frequency, and Energy

3.19 Which one of the following statements is TRUE?

a. The product of wavelength and frequency is a constant.
b. As the energy increases, the frequency of the radiation decreases.
c. As the wavelength of light increases, the frequency increases.
d. Red light has a higher frequency than blue light.
e. Light is only considered to have wave character.

70% 3.17 (b) In solving this problem, we use the same approach as we use in determining the empirical formula of a compound.

Divide each charge by the smallest charge.

$$\frac{-2.40 \times 10^{-19}}{-2.40 \times 10^{-19}} = 1.00$$

$$\frac{-3.20 \times 10^{-19}}{-2.40 \times 10^{-19}} = 1.33$$

$$\frac{-5.60 \times 10^{-19}}{-2.40 \times 10^{-19}} = 2.33$$

$$\frac{-6.40 \times 10^{-19}}{-2.40 \times 10^{-19}} = 2.67$$

$$\frac{-7.20 \times 10^{-19}}{-2.40 \times 10^{-19}} = 3.00$$

Now multiply each number by 3 to obtain all whole numbers and the number of electrons per drop.

76% 3.18 (b)

72% 3.19 (a) $\nu = \frac{c}{\lambda} \quad \therefore \nu\lambda = c$

3.20 What is the frequency of light having a wavelength of 4.50×10^{-6} cm? (The speed of light is 3.00×10^{8} m/s.) <u>Watch the units</u>; be consistent.

a. 1.50×10^{-16} s^{-1} b. 1.50×10^{-14} s^{-1} c. 1.34×10^{4} s^{-1}
d. 6.67×10^{13} s^{-1} e. 6.67×10^{15} s^{-1}

3.21 What is the wavelength of light in cm having a frequency of 1.5×10^{18} s^{-1}? The speed of light is 3.00×10^{8} m/s.

a. 2.0×10^{-10} b. 2.0×10^{-8} c. 5.0×10^{9} d. 5.0×10^{11}
e. 4.5×10^{28}

3.22 When a boron compound is heated in a flame, light with an energy content of 3.85×10^{-19} joules/quantum is emitted. Planck's constant is 6.67×10^{-34} joules x sec. From these facts and the diagram below, what color will be observed in the flame?

7.5×10^{14} 6.0×10^{14} 5.0×10^{14} 4.3×10^{14}

ν, sec^{-1}	Violet	Blue	Green	Yellow	Orange	Red

a. Blue to Blue Violet b. Green c. Yellow d. Orange e. Red

Modern Atomic Theory

3.23 The Rydberg constant expressed in units of radiation frequency is 3.293×10^{15} sec^{-1}. An electron in an excited hydrogen atom goes from a n=3 Bohr orbit to a n=2 Bohr orbit. The observation expected is

a. radiation of frequency 4.573×10^{14} sec^{-1} would be emitted.
b. radiation of frequency 3.293×10^{15} sec^{-1} would be emitted.
c. radiation of frequency 4.573×10^{14} sec^{-1} would be absorbed.
d. radiation of frequency 2.470×10^{15} sec^{-1} would be absorbed.
e. radiation of frequency 2.470×10^{15} sec^{-1} would be emitted.

92% 3.20 (e) $\nu = \frac{c}{\lambda} = \frac{3.00 \times 10^{8}\ \text{m/s} \times 100\ \text{cm/m}}{4.50 \times 10^{-6}\ \text{cm}}$

40% 3.21 (b) Be careful - this question asks for the wavelength of light in cm but the speed of light is given in m/s.

$$\lambda = \frac{c}{\nu} = \frac{3.00 \times 10^{8}\ \text{m/s} \times 100\ \text{cm/m}}{1.5 \times 10^{18}\ \text{s}^{-1}}$$

3.22 (b) $E = h\nu$

$$3.85 \times 10^{-19}\ \text{J} = 6.67 \times 10^{-34}\ \text{J·s}(\nu)$$

$$\nu = 5.77 \times 10^{14}\ \text{s}^{-1}$$

72% 3.23 (a) Since the electron is going from a higher energy level to a lower one, energy must be emitted.

$$\nu = R\left(\frac{1}{n_1^2} - \frac{1}{n_2^2}\right)$$

$$\nu = 3.293 \times 10^{15}\ \text{sec}^{-1}\left(\frac{1}{2^2} - \frac{1}{3^2}\right)$$

3.24 The maximum number of electrons that can occupy a single orbital is

a. 1 b. 2 c. 6 d. 8 e. 18

3.25 All the following are allowed energy levels EXCEPT

a. 3f b. 1s c. 3d d. 5p e. 6s

3.26 The number of orbitals in an f (ℓ=3) sub energy level is

a. 1 b. 3 c. 5 d. 7 e. 14

3.27 The maximum number of electrons that can exist in the n=4 energy level is

a. 4 b. 8 c. 16 d. 32 e. 64

3.28 When the principal quantum number n has the value 3, the azimuthal quantum number ℓ can have the values

a. +1/2, -1/2 b. 0, 1, 2 c. 0, 1, 2, 3 d. 2, 1, 0, -1, -2
e. 3, 2, 1, 0, -1, -2, -3

3.29 How many orbitals are there in a sublevel if the ℓ quantum number is equal to 3?

a. 1 b. 3 c. 5 d. 7 e. 9

3.30 Which one of the following electron cloud patterns is associated with a 2p orbital in modern wave mechanical (quantum mechanical) theory?

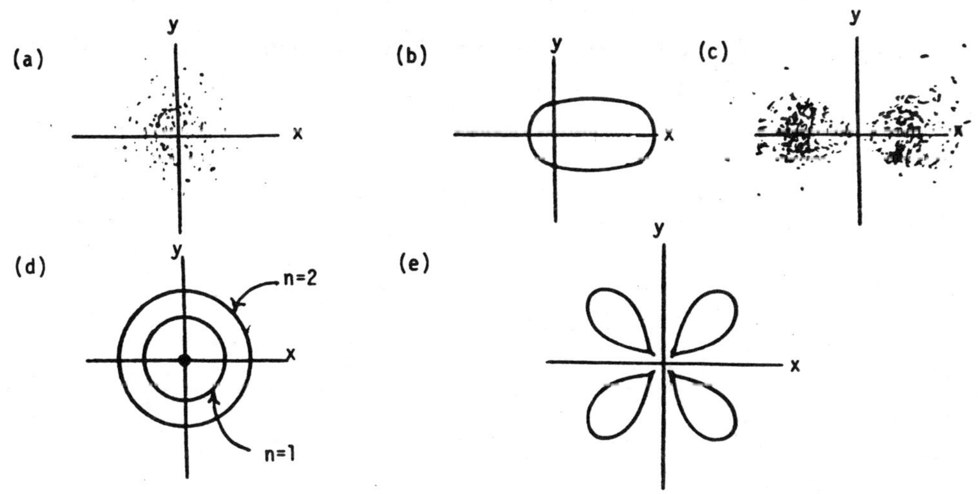

96% 3.24 (b) The electrons must have different spins.

96% 3.25 (a) The f sub energy level begins with 4f.

80% 3.26 (d) When $\ell = 3$, m_ℓ has the following values:
-3, -2, -1, 0, 1, 2, 3
This corresponds to seven orbitals.

3.27 (d) $2n^2 = 2(4)^2 = 32$

3.28 (b) ℓ has values from zero to n-1. Therefore, we get 0, 1 and 2.

3.29 (d) The ℓ quantum number of 3 corresponds to the f orbitals.

60% 3.30 (c)

3.31 The maximum number of electrons that can occupy the 5d orbitals is

a. 5 b. 10 c. 14 d. 18 e. 25

3.32 The ℓ quantum number represents

a. the shape of an orbital.
b. the number of valence electrons.
c. the magnetic properties of an electron.
d. the spin properties of an electron.
e. the angular momentum of an electron.

3.33 The two quantum numbers which represent a 3d electron are

a. $n = 3, \ell = 4$ b. $n = 3, \ell = 3$ c. $n = 4, \ell = 3$
d. $n = 3, \ell = 2$ e. $n = 3, \ell = 1$

3.34 Which of the following sets of quantum numbers is NOT allowed?

a. $n = 1, \ell = 0, m_\ell = 0, m_s = -1/2$
b. $n = 3, \ell = 1, m_\ell = -1, m_s = +1/2$
c. $n = 3, \ell = 3, m_\ell = -3, m_s = +1/2$
d. $n = 4, \ell = 3, m_\ell = 3, m_s = -1/2$
e. $n = 2, \ell = 0, m_\ell = 0, m_s = -1/2$

3.35 One of the outermost electrons in a calcium atom in the ground state can be described by which of the following sets of four quantum numbers?

a. 4, 2, 0, 1/2 b. 4, 1, 1, -1/2 c. 4, 1, 1, 1/2
d. 4, 0, 1, 1/2 e. 4, 0, 0, 1/2

3.31 (b)

88% 3.32 (a)

88% 3.33 (d)

90% 3.34 (c) ℓ must be at least one less than the n value.

84% 3.35 (e)

n	ℓ	m_ℓ	m_s
4	0	0	1/2

Calcium is $4s^2$, $\ell = 0$ for the s orbital. The m_ℓ value cannot be greater than the ℓ value.

3.36 Of the quantum state designations listed below, which do not describe an allowed state for an electron in an atom?

I. $n = 3,\ \ell = 2,\ m_\ell = -1$

II. $n = 3,\ \ell = 1,\ m_\ell = -1$

III. $n = 3,\ \ell = 0,\ m_\ell = -2$

IV. $n = 3,\ \ell = 2,\ m_\ell = 1$

V. $n = 3,\ \ell = 3,\ m_\ell = 0$

a. I and V b. II and III c. III and IV d. I and IV
e. III and V

3.37 Which of the following electron configurations represents a paramagnetic atom?

	1s	2s	2p
a.	↑↓	↑↓	↑ __ __
b.	↑↓	↑↓	↑ ↑ ↑
c.	↑↓	↑↓	↑↓ ↑ ↑

d. None of the above.
e. All of the above.

3.38 Which of the following electron configurations represents a diamagnetic atom?

	1s	2s	2p
a.	↑↓	↑↓	__ __ __
b.	↑↓	↑↓	↑ __ __
c.	↑↓	↑↓	↑ ↑ __
d.	↑↓	↑↓	↑ ↑ ↑
e.	↑↓	↑↓	↑↓ ↑ ↑

3.36 (e)

80% 3.37 (e) Atoms are paramagnetic when there are unpaired electrons. Boron, nitrogen, and oxygen <u>atoms</u> are all paramagnetic.

88% 3.38 (a) All the electrons are paired in diamagnetic atoms or molecules.

3.39 Which ground-state electronic configuration is INCORRECT?

a. $_{12}Mg$: $1s^2 2s^2 2p^6 3s^2$ b. $_{19}K$: $[Ar]4s^1$ c. $_{25}Mn$: $[Ar]4s^2 3d^4$

d. $_{17}Cl$: $[Ne]3s^2 3p^5$ e. $_{36}Kr$: $[Ar]4s^2 3d^{10} 4p^6$

3.40 The electronic configuration of the Al^{3+} ion is

a. $1s^2 2s^2 2p^6$ b. $1s^2 2s^2 2p^6 3s^2 3p^1$ c. $1s^2 2s^2 2p^6 3s^2 3p^4$

d. $1s^2 2s^2 2p^3$ e. $1s^2 2s^2 2p^1$

3.41 The Fe atom would be expected to have ___ unpaired electrons.

a. 6 b. 5 c. 4 d. 3 e. 2

3.42 From a consideration of electronic configurations, which of the elements indicated below would be classified as a transition element?

a. $1s^2$, $2s^2$, $2p^2$ b. $1s^2$, $2s^2$, $2p^6$, $3s^2$, $3p^5$

c. $1s^2$, $2s^2$, $2p^6$, $3s^2$, $3p^6$, $4s^2$ d. $1s^2$, $2s^2$, $2p^6$, $3s^2$, $3p^6$ $3d^5$, $4s^2$

e. $1s^2$, $2s^2$, $2p^6$, $3s^2$, $3p^6$, $3d^{10}$, $4s^2$, $4p^6$

3.43 The number of unpaired electrons in an atom of arsenic (element No. 33) is

a. 0 b. 1 c. 2 d. 3 e. 5

3.44 If each of the following species were in their lowest energy states, which set contains species which have the same electronic configuration?

a. F^-, Cl^- Br^-, I^- b. Na, Mg, Al, Si c. Li, Na, K, Rb

d. Na^+, Cl^-, Ne , F e. S^{2-}, Cl^-, K^+, Ar

96% 3.39 (c) In the electronic configuration for Mn, there are shown 24 electrons not 25.

80% 3.40 (a) Al^{3+} ion has 10 electrons, the neon configuration.

3.41 (c) Iron has the electronic configuration:
$[Ar]\ 4s^2\ 3d^6$

↑↓ ↑ ↑ ↑ ↑
3d

92% 3.42 (d) A transition element is one in which there is an incomplete d sub energy level.

3.43 (d) Arsenic : $[Ar]\ 4s^2\ 3d^{10}\ 4p^3$

↑ ↑ ↑
4p

88% 3.44 (e) Which set has the same number of electrons? Which set is isoelectronic?

Ionization Energy and Electron Affinity

3.45 Which of the following processes represents the ionization energy of bromine?

a. $Br_2(\ell) \rightarrow Br_2^+(\ell) + e$ b. $Br(\ell) \rightarrow Br^+(\ell) + e$

c. $Br(\ell) \rightarrow Br^+(g) + e$ d. $Br(g) \rightarrow Br^+(g) + e$

e. $Br(g) + e \rightarrow Br^-(g)$

3.46 Which of the following elements has the SMALLEST first ionization energy?

a. $_{15}P$ b. $_{19}K$ c. $_{6}C$ d. $_{12}Mg$ e. $_{18}Ar$

3.47 Which of the following has the highest ionization energy for the removal of the <u>third</u> electron?

a. Mg b. Al c. Si d. Cl e. Ar

3.48 Of the following, the atom with the largest first ionization energy is the one with the electronic configuration

a. $1s^2\ 2s^2\ 2p^2$ b. $1s^2\ 2s^2\ 2p^3$ c. $1s^2\ 2s^2\ 2p^6\ 3s^2\ 3p^2$

d. $1s^2\ 2s^2\ 2p^6\ 3s^2\ 3p^3$ e. $1s^2\ 2s^2\ 2p^6\ 3s^2\ 3p^4$

3.45 (d) The ionization energy is defined as the energy required to remove an electron from the gaseous atom to produce the gaseous ion.

76% 3.46 (b) Potassium is a large atom which is at the beginning of a period where a new higher energy level is starting to fill.

68% 3.47 (a) Magnesium has only two valence electrons. All the other atoms in this question have three or more valence electrons. To remove a third electron from magnesium, it must come from a lower energy level.

3.48 (b) The atom which is the smallest usually has the highest ionization energy. Elements in the upper right hand corner of the Periodic Table will have the highest ionization energy.

The data and statements below are to be used for questions 3.49 and 3.50.

Element	First Ionization Energy (kcal/mole)
H	314
He	567
Li	124
Be	215
B	191
C	260
N	335
O	314
F	402
Ne	497

I. An increase in nuclear charge.

II. A screening effect by the inner electrons.

III. p-electrons are slightly higher in energy than s-electrons of the same principal quantum number.

IV. Extra repulsion results from two electrons occupying the same orbital.

3.49 The low ionization energy of lithium as compared to helium can best be accounted for by

a. I b. II c. IV d. I and II e. II and IV

3.50 The fact that the ionization energy of oxygen is much greater than beryllium and less than nitrogen can best be accounted for by

a. I and II b. I and III c. I and IV d. II and III e. II and IV

3.51 For the isoelectronic series, which is given below, the atom or ion which would require the least energy for the removal of an electron is

a. $_{15}P^{3-}$ b. $_{16}S^{2-}$ c. $_{17}Cl^{-}$ d. $_{18}Ar$ e. $_{19}K^{+}$

3.49 (b) The valence electron of lithium is in a higher energy level. This electron is screened from the nucleus by the $1s^2$ electrons.

3.50 (c) Both Be and O are in the second period but O has four more positive charges and thus is a smaller atom. Consequently it takes more energy to remove an electron from oxygen.

N ↑↓ (2s) ↑ ↑ ↑ (2p)

O ↑↓ (2s) ↑↓ ↑ ↑ (2p)

Even though O has one more proton than N, the repulsion resulting from two electrons in the 2 p orbital makes the first ionization energy of O less than that of N.

3.51 (a) It takes less energy to remove an electron from species that has more electrons than protons. Also select a species that has the fewest protons.

3.52 The first ionization energy of cesium is 6.24×10^{-19} J/atom. Calculate the minimum frequency of light that is required to ionize a cesium atom. The value for Planck's constant is 6.626×10^{-34} J•s.

a. $1.06 \times 10^{-15}\ s^{-1}$ b. $1.06 \times 10^{-12}\ s^{-1}$ c. $9.42 \times 10^{11}\ s^{-1}$

d. $4.13 \times 10^{14}\ s^{-1}$ e. $9.42 \times 10^{14}\ s^{-1}$

3.53 Which <u>one</u> of the following atoms should have the largest value for the <u>electron affinity</u> of the free atom? Atoms are designated by electronic configuration.

a. $1s^2\ 2s^2\ 2p^6\ 3s^1$ b. $1s^2\ 2s^2\ 2p^5$ c. $1s^2\ 2s^2\ 2p^6\ 3s^2\ 3p^2$

d. $1s^2\ 2s^2\ 2p^6\ 3s^2\ 3p^6\ 3d^5\ 4s^2$ e. $1s^2\ 2s^2\ 2p^6$

Atomic and Ionic Radii

3.54 The elements in which of the following sequences have most nearly the same atomic radius?

a. Mg, Al, Si, P b. F, Cl, Br, I c. Li, Na, K, Rb

d. O, S, Cl, Ar e. Mo, Tc, Ru, Rh

3.55 The correct order for <u>decreasing</u> atomic radius of the five atoms Cs, Ba, Cl, Br, I is

a. I > Br > Cl > Cs > Ba b. Ba > Cs > I > Br > Cl

c. Cs > Ba > I > Br > Cl d. Ba > Cs > Cl > Br > I

e. Cl > I > Br > Ba > Cs

3.56 Which atom has the SMALLEST atomic radius?

a. $_{19}K$ b. $_{31}Ga$ c. $_{35}Br$ d. $_{37}Rb$ e. $_{52}Te$

3.57 Which one of the following has the SMALLEST atomic radius?

a. Li b. Cs c. F d. Na e. Cl

88% 3.52 (e) $E = h\nu \; ; \; \nu = \dfrac{6.24 \times 10^{-19}\ J}{6.626 \times 10^{-34}\ J\cdot s}$

3.53 (b) The electronic configuration which corresponds to the fluorine atom has the greatest electron affinity. Fluorine is a small atom with a relatively large nuclear charge and thus a large attraction for electrons.

68% 3.54 (e) The radii of the transition elements in a given period do not change very much because an inner energy level is being filled.

3.55 (c) The alkali metals are larger than the corresponding halogens. The radii increase down a given family of elements.

92% 3.56 (c) Going from left to right across the Periodic Table in a given period, there is a general decrease in size because of an increase in nuclear charge. Going from top to bottom in a family or group there is a general increase because of an increase in the number of energy levels.

3.57 (c) The smallest atoms are at the top of a family and to the right-hand side of a period. Why?

3.58 Which of the following is NOT in the order of increasing ionic radius?

a. $Ca^{2+} < Sr^{2+} < Ba^{2+}$ b. $Al^{3+} < Mg^{2+} < Na^{+}$ c. $K^{+} < Cl^{-} < S^{2-}$

d. $O^{2-} < F^{-} < Na^{+}$ e. $Cl^{-} < Br^{-} < I^{-}$

3.59 Which of the following ions has the SMALLEST radius?

a. Mg^{2+} b. Al^{3+} c. P^{3-} d. S^{2-} e. Cl^{-}

3.60 The radii of the species O, O^{+} and O^{-} decrease in the following order

a. $O>O^{+}>O^{-}$ b. $O>O^{-}>O^{+}$ c. $O^{+}>O^{-}>O$ d. $O^{-}>O>O^{+}$

e. $O^{-}>O^{+}>O$

Oxidation Numbers

3.61 The oxidation number of boron in the compound H_3BO_3 is

a. 0 b. 1 c. 2 d. 3 e. 6

3.62 The oxidation number of C in CH_2O is

a. -4 b. -2 c. 0 d. +2 e. +4

3.63 The oxidation number of nitrogen in sodium nitrite, $NaNO_2$, is

a. -1 b. +1 c. +2 d. +3 e. +4

3.64 The oxidation number of phosphorus in $CaHPO_4$ is

a. -2 b. +1 c. +3 d. +4 e. +5

3.65 The oxidation number of Cr in $Na_2Cr_2O_7$ is

a. +3 b. +4 c. +5 d. +6 e. +7

3.66 The oxidation number of Tc in $KTcO_4$ is

a. 3 b. 4 c. 5 d. 6 e. 7

60% 3.58 (d) Na^+ would be much smaller than O^{2-} because there are 11 protons attracting 10 electrons whereas O^{2-} has 8 protons attracting 10 electrons.

3.59 (b) Positive ions are usually smaller than negative ions. Why? Both Mg^{2+} and Al^{3+} are isoelectronic but Al^{3+} has one more proton attracting the same number of electrons.

3.60 (d) For species with the same number of protons, the one with the most electrons will be the largest and the one with the fewest electrons will be the smallest. Why?

3.61 (d) Hydrogen bonded to nonmetals has an oxidation number of +1.

$3(+1) + y + 3(-2) = 0$

3.62 (c) Yes, there is an oxidation number of zero for an atom in a compound.

$y + 2(+1) + 1(-2) = 0$

84% 3.63 (d) NO_2^-, $y + 2(-2) = -1$

84% 3.64 (e) HPO_4^{2-}, $+1 + y + 4(-2) = -2$

80% 3.65 (d) $Cr_2O_7^{2-}$, $2y + 7(-2) = -2$

90% 3.66 (e) TcO_4^-, $y + 4(-2) = -1$

4

BONDING

Ions and Formulas of Ionic Compounds

4.1 How many electrons does a stable aluminum ion have?

a. 10 b. 11 c. 12 d. 13 e. 16

4.2 What is the electronic configuration of a stable magnesium ion?

a. $1s^2\ 2s^2\ 2p^5$ b. $1s^2\ 2s^2\ 2p^6$ c. $1s^2\ 2s^2\ 2p^6\ 3s^1$

d. $1s^2\ 2s^2\ 2p^6\ 3s^2$ e. $1s^2\ 2s^2\ 2p^6\ 3s^2\ 3p^1$

4.3 All of the following represent stable positive ions EXCEPT

a. Ba^{2+} b. Ga^{3+} c. Na^+ d. Ca^+ e. Pb^{4+}

4.4 Which of the following represents a known stable ion?

a. O^- b. Li^{2+} c. Cl^{2-} d. F^+ e. S^{2-}

4.5 An element, X, has the electronic configuration $1s^2 2s^2 2p^3$. The formula of the most probable compound this element will form with magnesium is

a. MgX b. MgX_2 c. MgX_3 d. Mg_3X_2 e. Mg_2X_3

4.6 An element X from Group IIA reacts with Y from Group VA. Which is true concerning the elements or compound?

a. X is a nonmetal. b. Y is a metal.

c. The compound's formula is X_3Y_2. d. The compound would be covalent.

e. Elemental properties are preserved in the compound.

4.1 (a) One needs to know the formulas of some common ions. Al is in family III, therefore it is Al^{3+}. 13-3 = 10

4.2 (b) Magnesium is in family II, therefore, the magnesium ion is Mg^{2+}, 10 electrons.

4.3 (d) Calcium is in family II, so we would expect the calcium ion to be 2+ so that it has a noble gas configuration.

4.4 (e) Simple stable negative ions are also expected to have noble gas configurations.

88% 4.5 (d) X is from family V, so it would be X^{3-} as a simple negative ion.

4.6 (c) See question 4.5 above.

4.7 $1s^2 2s^2 2p^6 3s^2 3p^4$

Atoms of an element X have the electronic configuration shown above. The compound most likely formed with aluminum is

a. AlX b. AlX_2 c. Al_2X_3 d. Al_3X_2 e. Al_2X

4.8 The formulas of the nitrite, nitride, and nitrate ions are represented respectively as

a. NO_2^-, NO_3^-, NO_4^- b. N^{3-}, NO_2^-, NO_3^- c. NO^-, NO_2^-, NO_3^-

d. NO_2^-, N^{3-}, NO_3^- e. NO_3^-, NO_2^-, N^{3-}

4.9 The formulas of the hydroxide ion, the nitrate ion and the phosphate ion are represented respectively as

a. OH^-, NO_3^-, PO_4^{3-} b. H^-, NO_2^-, P^{3-} c. H^-, NO_3^-, P^{3-}

d. OH^-, NO_2^-, PO_3^{3-} e. OH^-, NO_2^-, PO_4^{3-}

4.10 The formulas of the phosphate ion, the ammonium ion, and the chlorate ion are represented respectively as

a. PO_4^{3-}, NH_4^+, ClO_3^- b. PO_3^{3-}, NH_2^-, ClO_4^- c. P^{3-}, NH_2^+, ClO_2^-

d. PO_2^-, NH_3^+, ClO^- e. P^{3-}, NH_4^+, Cl^-

4.11 The formula for aluminum bromide is

a. AlB b. AlBr c. Al_2Br_3 d. $AlBr_2$ e. $AlBr_3$

4.12 What formula would be expected for a binary compound of calcium and nitrogen?

a. CaN b. Ca_3N_2 c. Ca_2N d. Ca_2N_3 e. Ca_5N_2

84% 4.7 (c) Element X is from family VI, so the simple ion will be X^{2-}.

64% 4.8 (d)

4.9 (a) A student should have the names and formulas of these ions memorized.

92% 4.10 (a) See the comment to question 4.9.

4.11 (e) Al^{3+} and Br^{-}

4.12 (b) Ca^{2+} and N^{3-}

4.13 The formula for terbium phosphate is $Tb_3(PO_4)_4$. Based upon this information, the formula for the sulfate of terbium would be expected to be

a. $Tb_2(SO_3)_3$ b. Tb_2S_3 c. TbS_2 d. $Tb_2(SO_4)_3$

e. $Tb(SO_4)_2$

4.14 The formula of praseodymium nitrate is $Pr(NO_3)_4$. Based upon this information, the formula for the phosphate of praseodymium would be expected to be

a. PrPo b. $PrPO_4$ c. $Pr_3(PO_4)_4$ d. $Pr_2(PO_4)_3$ e. Pr_2P_3

4.15 The formula for holmium carbonate is $Ho_2(CO_3)_3$. Based upon this information, the formula for the nitrate of holmium would be expected to be

a. $Ho(NO_3)_2$ b. $Ho(NO_3)_3$ c. Ho_2N_3 d. Ho_3N_2 e. $HoNO_3$

4.16 The formula of potassium carbonate is

a. P_2CO_3 b. K_2CO_3 c. Po_2CO_3 d. P_2C e. K_2C

4.17 The formula of potassium sulfite is

a. P_2SO_4 b. K_2SO_4 c. K_2S d. KSO_4 e. K_2SO_3

4.18 What value does n have in the BeF_4^n ion?

a. -4 b. -2 c. -1 d. 0 e. +2

4.19 What value does n have in the SiF_6^n ion?

a. -4 b. -2 c. 0 d. +2 e. +4

72% 4.13 (e) The phosphate ion is PO_4^{3-}, so terbium must be Tb^{4+}. The sulfate ion is SO_4^{2-}.

4.14 (c) NO_3^-, Pr^{4+}, PO_4^{3-}

80% 4.15 (b) The carbonate ion is CO_3^{2-}, so holmium must be 3+. Since the nitrate ion has a minus one charge, three nitrate ions are required to balance the charge of the holmium ion.

4.16 (b) K^+ amd CO_3^{2-}. Remember, the potassium ion is K^+. The symbol for phosphorus is P.

92% 4.17 (e) Potassium is K^+ and the sulfite ion is SO_3^{2-}.

4.18 (b) Beryllium is from family II, so we should have Be^{2+}. The fluoride ion is F^-, so $Be^{2+} + 4F^- = 2$ minus.

80% 4.19 (b) Silicon is in family IV, so Si would have a 4^+ oxidation number. $Si^{4+} + 6F^- = 2^-$

4.20 What value would you predict for n in the PCl_4^n ion?

a. -4 b. -1 c. 0 d. +1 e. +3

Ionic and Covalent Bonding

4.21 Which of these compounds has the greatest ionic character?

a. ClF b. IF c. CsF d. AlF_3 e. CF_4

4.22 In which of the following is there <u>both</u> ionic <u>and</u> covalent bonding?

a. NH_4Cl b. $Ba(ClO_3)_2$ c. $Ba(ClO)_2$ d. $Ca(NO_2)_2$

e. All of these.

4.23 Which compound contains both ionic and covalent bonds?

a. NaCN b. NaCl c. HCl d. CO_2 e. SO_2

4.24 Which compound contains both ionic and covalent bonds?

a. NH_4NO_3 b. NH_3 c. N_2O_5 d. NaI e. NO

4.25 Which compound contains only ionic bonding?

a. NaCN b. KSCN c. $CsNO_3$ d. LiBr e. $MgCO_3$

40% 4.20 (d) Phosphorus is in family V, so we could expect it to have an oxidation number of 5^+. $5^+ + 4Cl^-$ = plus one.

4.21 (c) Cesium is the most electro-positive atom and fluorine is the most electronegative atom.

4.22 (e) All the compounds shown are salts. One ion in each salt is polyatomic with covalent bonding holding the atoms together.

96% 4.23 (a) $Na^+ + CN^-$. The carbon atom and nitrogen atom in the cyanide ion are sharing bonding electrons.

4.24 (a) $NH_4^+ + NO_3^-$ See answer to question 4.23.

92% 4.25 (d) $Li^+ + Br^-$ are simple ions that form a crystal lattice by electrostatic interactions between the charges on the ions.

Ionic Attractions (Crystal Lattice Energies and Melting Points)

The diagram below will be used for questions 4.26 and 4.27

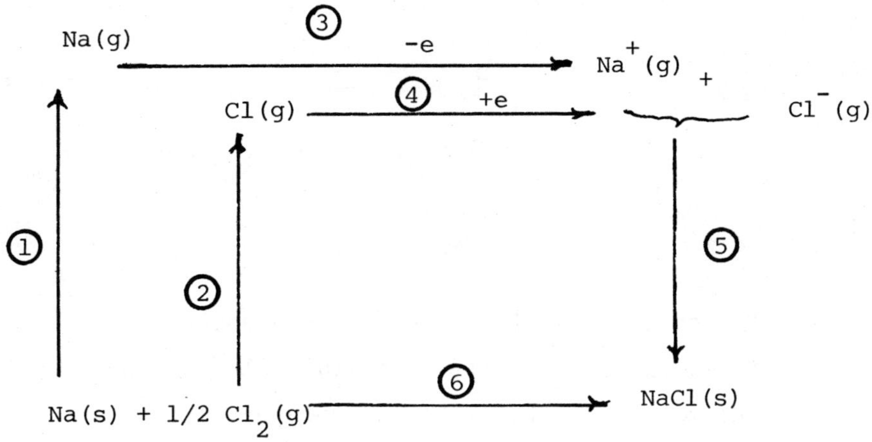

4.26 The crystal lattice energy is represented by step(s)

a. 1   b. 5   c. 6   d. 3, 4, 5   e. 1, 2, 3, 4, 5

4.27 The heat of formation of sodium chloride can be found from step(s)

a. 1 + 2   b. 2 + 4   c. 5 - 6   d. 3 + 4 + 5

e. 1 + 2 + 3 + 4 + 5

70% 4.26 (b) The crystal lattice energy is the energy given off in the formation of one mole of solid NaCl from the combination of one mole of $Na^+(g)$ with one mole of $Cl^-(g)$.

4.27 (e) ΔH_f NaCl = $\Delta H_{sublimation}$ + 1/2 $\Delta H_{diss.}(Cl_2)$ + I.E. (Na) + E.A. (Cl) + U(crystal lattice energy)

Reaction	Step
$Na(s) \rightarrow Na(g)$	1
$1/2\ Cl_2(g) \rightarrow Cl(g)$	2
$Na(g) \rightarrow Na^+(g) + e$	3
$Cl(g) + e \rightarrow Cl^-(g)$	4
$Na^+(g) + Cl^-(g) \rightarrow NaCl(s)$	5
$Na(s) + 1/2\ Cl_2(g) \rightarrow NaCl(s)$	1+2+3+4+5

Questions 4.28 - 4.30 consist of five letter headings which are followed by the three questions. For each question, select the one heading which is most closely related to it.

a. $Li(s) + 1/2\ Cl_2(g) \rightarrow LiCl(s)$

b. $Cl^-(g) \rightarrow Cl(g) + e$

c. $Li^+(g) + Cl^-(g) \rightarrow LiCl(s)$

d. $Li(g) \rightarrow Li(s)$

e. $Li^+(g) + Cl^-(g) \rightarrow LiCl(g)$

4.28 Is an endothermic process.

4.29 U, crystal lattice energy.

4.30 ΔH, heat of formation.

The following thermochemical equations are to be used for questions 4.31 and 4.32.

$$1/2\ X_2(g) \rightarrow X(g) \qquad \Delta H = 30 \text{ kcal/mole}$$
$$M(s) \rightarrow M(g) \qquad \Delta H = 22 \text{ kcal/mole}$$
$$M(g) \rightarrow M^+(g) + e \qquad \Delta H = 118 \text{ kcal/mole}$$
$$M^+(g) + X^-(g) \rightarrow MX(s) \qquad \Delta H = -190 \text{ kcal/mole}$$
$$M(s) + 1/2\ X_2(g) \rightarrow MX(s) \qquad \Delta H = -100 \text{ kcal/mole}$$

4.31 The value for ΔH formation of MX(s) is

a. -190 b. 190 c. -100 d. 100

e. cannot be determined from the data above.

4.32 Calculate ΔH for the electron affinity of X(g) ($X(g) + e \rightarrow X^-(g)$)

a. -80 kcal b. -30 kcal c. -20 kcal d. 20 kcal e. 156 kcal

4.33 Which of the following compounds would be expected to have the largest crystal lattice energy?

a. NaF b. NaCl c. NaBr d. NaI e. KI

60% 4.28 (b) $Cl(g) + e \rightarrow Cl^-(g)$, corresponds to the electron affinity which is exothermic. The reverse of the above process must be endothermic.

68% 4.29 (c) See question 4.26.

84% 4.30 (a)

52% 4.31 (c) The heat of formation of a compound corresponds to the energy change that occurs when the compound is formed from the elements in their most stable form at 25°C.

52% 4.32 (a) $E.A. = -\Delta H\ (M)_{sublimation} - 1/2\ \Delta H\ (X_2)_{dissociation} - I.E. - U + \Delta H_f$

$= -22 - 30 - 118 + 190 - 100 = -80$ kcal

	ΔH, kcal
$M(g) \rightarrow M(s)$	-22
$X(g) \rightarrow 1/2\ X_2(g)$	-30
$M^+(g) + e \rightarrow M(g)$	-118
$MX(s) \rightarrow M^+(g) + X^-(g)$	+190
$M(s) + 1/2\ X_2(g) \rightarrow MX(s)$	-100
$X(g) + e \rightarrow X^-(g)$	-80 kcal

4.33 (a) The crystal lattice energy, U is proportional to the charges on the ions and inversely proportional to the size of the ions or distance between the ions.

$$U \alpha \frac{q_1 q_2}{r_o}$$

4.34 Which of these compounds would be expected to have the largest crystal lattice energy?

a. NaCl b. $MgCl_2$ c. KCl d. CCl_4 e. $SiCl_4$

435. ____________ would be expected to have the highest melting point.

a. KI b. KF c. NaCl d. NaF e. NaI

4.36 Which of these compounds would be expected to have the highest melting point?

a. NaF b. CsI c. MgO d. BF_3 e. CF_4

Covalent Bonding

4.37 A covalent bond is the result of the

a. overlap of two half-filled s orbitals.
b. overlap of a half-filled s orbital and a half-filled p orbital.
c. overlap of two half-filled p orbitals along their axes.
d. sidewise overlap of two half-filled parallel p orbitals.
e. All of the above.

4.38 Which is the best definition of a covalent bond?

a. Electrons simultaneously attracted by more than one nucleus.
b. Overlapping of two electron-filled orbitals having different energies.
c. Overlapping of unoccupied orbitals of two or more atoms.
d. Positive ion attracting negative ions.
e. Interaction between outer electrons.

30% 4.34 (b) CCl_4 and $SiCl_4$ are covalent compounds. $MgCl_2$ has the largest crystal lattice energy because magnesium has a plus 2 charge which greatly increases the lattice energy. See question 4.33.

40% 4.35 (d) There is a correlation between the melting point and the crystal lattice energy. The larger the lattice energy the higher the melting point.

50% 4.36 (c) A plus two ion combining with a negative ion gives a very large crystal lattice energy. See questions 4.33 and 4.35.

4.37 (e) A covalent bond involves the sharing of bonding electrons.

4.38 (a) A sharing of electrons.

4.39 The formulas of many binary covalent compounds can be predicted on the basis

a. that a bond is formed by the overlapping of two filled orbitals.

b. that a bond is formed by the overlapping of atomic orbitals.

c. that the number of bonds that an atom can have is equal to the number of empty valence orbitals it has.

d. that the number of bonds is equal to the number of half-filled valence orbitals it can have.

e. that bonding electrons are simultaneously attracted by more than one nucleus.

4.40 The existence of SF_2, SF_4 and SF_6 can best be explained by

I. Each bond in the three compounds is formed by the overlapping of two half-filled orbitals.

II. The number of bonds that an atom can have is equal to the number of half-filled valence orbitals it has.

III. The number of bonds that an atom can have is equal to the number of half-filled orbitals it can have.

a. I only b. II only c. III only d. I and II only

e. I, II and III

4.41 Which formula represents the most probable oxygen compound?

a. NaO b. HOOH c. OF d. Ca_2O_3 e. AlO

4.42 Which formula represents the most probable bromine compound?

a. CaBr b. AlBr c. BrCl d. HBr_2 e. NBr_2

4.43 Which formula represents the most probable hydrogen compound?

a. BaH b. H_2F c. PH_4 d. SiH_4 e. BH_4

4.44 Which of the following atoms always violates the octet rule?

a. F b. P c. S d. H e. Cl

90% 4.39 (d) A covalent bond results from the overlap of two half-filled orbitals. For example:

F $2s^2 2p^5$ one bond , HF

O $2s^2 2p_x^2 p_y^1 p_z^1$ two bonds , H_2O

N $2s^2 2p_x^1 p_y^1 p_z^1$ three bonds , NH_3

40% 4.40 (e) S in SF_2 $3s^2 3p_x^2 p_y^1 p_z^1$

S in SF_4 $3s^2 3p_x^1 p_y^1 p_z^1 3d^1$

S in SF_6 $3s^1 3p_x^1 p_y^1 p_z^1 3d^1 d^1$

88% 4.41 (b) See question 4.39. oxygen has the tendency to form two covalent bonds because it has two half-filled orbitals. In distractors a, d and e, there is not a charge balance in these ionic type compounds.

4.42 (c) In covalent compounds, bromine and chlorine would have the tendency to form one bond as each has one half-filled orbital.

4.43 (d) Silicon would have the tendency to form four bonds.

4.44 (d) Since hydrogen is from the first period, the maximum number of electrons it can have is two.

4.45 Which of these compounds violates the octet rule for electron distribution around the central atom?

a. CO_2 b. NF_3 c. OF_2 d. PF_5 e. N_2

4.46 In which of the following compounds is the octet rule violated by one of the atoms in the molecule?

a. NO b. CO c. SO_2 d. SO_3 e. CF_4

4.47 In which of the following species is the octet rule violated by the central atom?

a. PCl_3 b. PCl_4^+ c. CCl_3^- d. BCl_3 e. NCl_3

Electron Counting

4.48 All the following species are isoelectronic with argon EXCEPT

a. Ca^{2+} b. K^+ c. S^{2-} d. Ga^{3+} e. Cl^-

4.49 Which pair of species is isoelectronic?

a. K^+ and Ca^{2+} b. Na^+ and K^+ c. Cl and Ar d. Cl^- and K

e. Br and I

4.50 All the following species are isoelectronic (have the same number of valence electrons) EXCEPT

a. ${}_{52}TeF_6$ b. ${}_{51}SbF_6^-$ c. ${}_{54}XeF_6$ d. ${}_{50}SnF_6^{2-}$

e. ${}_{49}InF_6^{3-}$

4.51 From a consideration of only the valence electrons, all the following are isoelectronic EXCEPT

a. SO_3 b. SeO_3 c. NO_3^- d. CO_3^{2-} e. IO_3^-

4.45 (d) 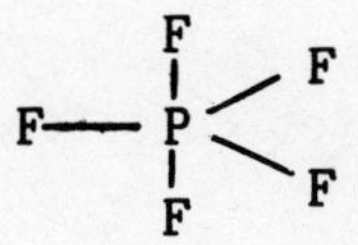

P has ten valence electrons around it.

4.46 (a) NO has an odd number of valence electrons (N (5) and O (6)), so it is impossible to have an even number for an octet.

4.47 (d) 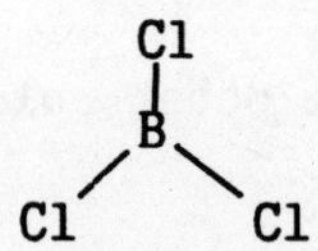

Boron has six valence electrons (from the three bonding pairs) around it.

4.48 (d) Ga^{3+} : [Ar] $3d^{10}$

4.49 (a) Both K^+ and Ca^{2+} have 18 electrons which is the argon configuration, ${}_{18}Ar$.

96% 4.50 (c) Each central atom has 6 bonds and 12 valence electrons about it except for Xe which has 6 bonds plus a lone pair of electrons about it.

4.51 (e) Draw the Lewis structures or count the number of valence electrons. The iodate ion has 26 valence electrons whereas all the other species in this question have 24 valence electrons.

4.52 The total number of valence electrons in the selenite ion, SeO_3^{2-}, is

a. 24 b. 26 c. 32 d. 54 e. 60

4.53 The number of valence electrons in the acetate ion, CH_3COO^- is

a. 22 b. 23 c. 24 d. 36 e. 38

4.54 The total number of valence electrons in the oxalate ion, $C_2O_4^{2-}$, is

a. 28 b. 30 c. 32 d. 34 e. 36

4.55 The total number of valence electrons in the perchlorate ion is

a. 24 b. 26 c. 28 d. 30 e. 32

4.56 The total number of valence electrons in the cyanide ion is

a. 8 b. 9 c. 10 d. 16 e. 24

Lewis Structures

4.57 Which of the following species does NOT have the proper Lewis formula?

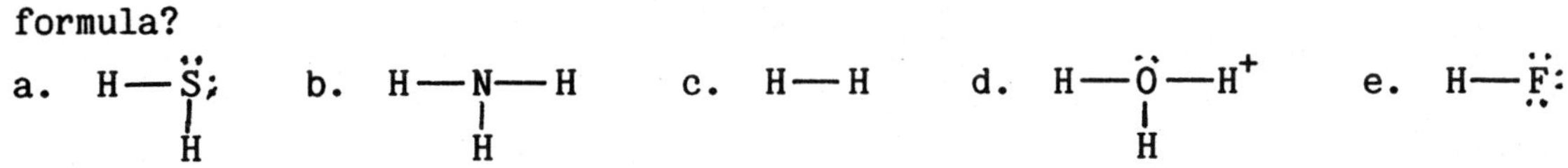

92% 4.52 (b)

Se : 6
3-O : 18
2e : 2
26

4.53 (c)

2-C : 8
2-O : 12
3-H : 3
1e : 1
24

4.54 (d)

2-C : 8
4-O : 24
2e : 2
34

4.55 (e) This problem involves two steps as it is necessary for a student to know the formula of the perchlorate ion, ClO_4^-.

Cl : 7
4-O : 24
1e : 1
32

4.56 (c) CN^-

C : 4
N : 5
1e : 1
10

4.57 (b) First count the number of valence electrons.

N : 5
3-H : 3
8

Therefore, the formulas for NH_3 is

```
     ..
 H - N - H
     H
```

4.58 Which of the following covalent molecules does not have the proper Lewis formula?

a. 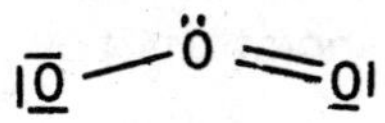

b. 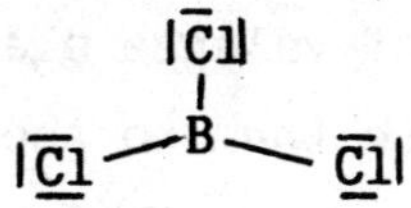

c. 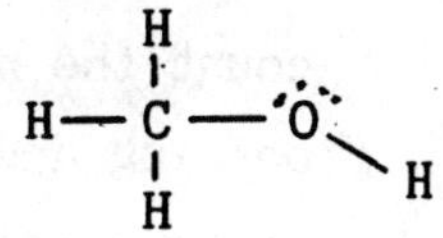

d. 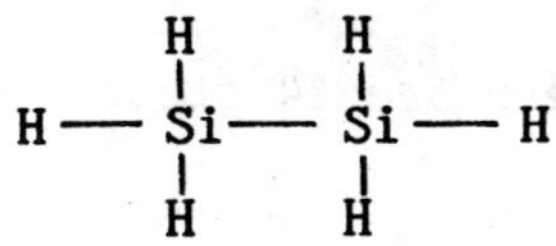

e. 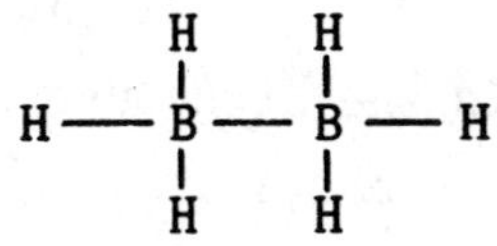

4.59 Which of the following species does not have the proper Lewis formula?

a. 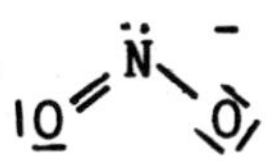

b. $|O = N = O|^{+}$

c. $|N \equiv O|^{+}$

d. 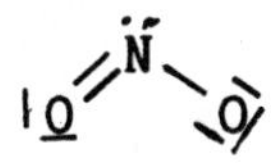

e. 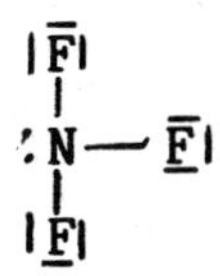

4.60 Which one of the following would have a Lewis structure(s) most like that of CO_3^{2-}?

a. NO_3^- b. NH_3 c. CH_3^+ d. SO_4^{2-} e. SO_3^{2-}

4.61 Which one of the following species would have a Lewis structure most like that of SO_2?

a. CO_2 b. NO_2 c. NO_2^+ d. NO_2^- e. SiO_2

72% 4.58 (e) First, make sure that each molecule has an octet of electrons around all atoms other than hydrogen. Second, count the number of valence electrons that are available for bonding and compare them to the number shown in the Lewis structures.

B_2H_6 2-B : 6
6-H : 6
12 electrons

In the structure shown in (e) there are 14 electrons (two too many!).

4.59 (d) NO_2 is shown with one extra electron.

4.60 (a) Find two species with the same number of valence electrons or draw the Lewis structures.

CO_3^{2-}		NO_3^-	
1-C :	4	1-N :	5
3-O :	18	3-O :	18
2e :	2	1e :	1
	24		24

4.61 (d)

SO_2		NO_2^-	
1-S :	6	1-N :	5
2-O :	12	2-O :	12
	18	1e :	1
			18

4.62 Which one of the following would have a Lewis structure(s) most like that of SO_3^{2-}?

a. SO_4^{2-} b. NO_3^- c. SO_3 d. NF_3 e. CO_3^{2-}

4.63 In the Lewis structure for the OF_2 molecule, the number of lone pairs of electrons around the central oxygen atom is

a. 0 b. 1 c. 2 d. 3 e. 4

4.64 The central atom in the bromite ion BrO_2^- is surrounded by

a. two bonding and two unshared pairs of electrons.
b. three bonding and one unshared pair of electrons.
c. one bonding and three unshared pairs of electrons.
d. two double bonds and no unshared pairs of electrons.
e. four bonding and four lone pairs of electrons.

4.65 In the Lewis electron dot structure for hydrazine, N_2H_4, the total number of lone electron pairs around the two nitrogen atoms is

a. 0 b. 1 c. 2 d. 3 e. 4

4.66 In the Lewis structure of hydrazoic acid, HNNN, the total number of lone pairs of electrons around the nitrogen atoms is

a. 0 b. 1 c. 2 d. 3 e. 4

4.67 Which one of the following species has a Lewis structure most like that of CO_2?

a. HCN b. SO_2 c. NO_2 d. NO_2^+ e. NO_2^-

4.68 In the Lewis structure for SF_4, the number of lone pairs of electrons around the central sulfur atom is

a. 0 b. 1 c. 2 d. 4 e. 5

80% 4.62 (d) It is best to draw the Lewis structures.

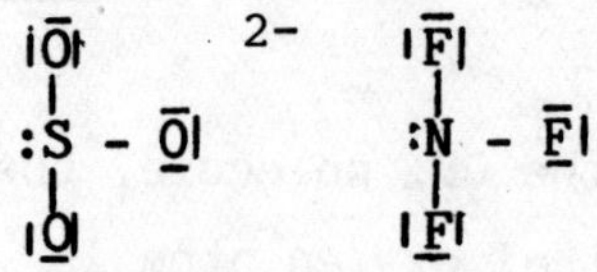

4.63 (c)

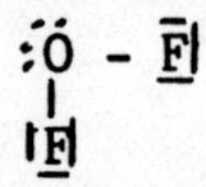

4.64 (a)

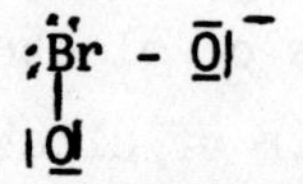

4.65 (c)

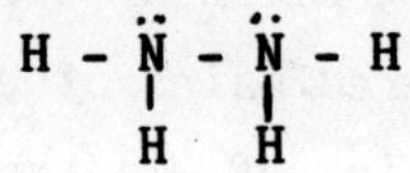

4.66 (d) <u># valence electrons</u>

H : 1

3-N : <u>15</u>

16

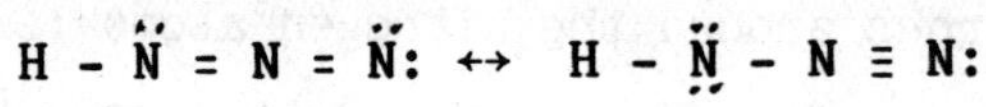

80% 4.67 (d)

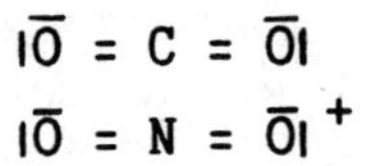

88% 4.68 (b) <u># valence electrons</u>

1-S : 6

4-F : <u>28</u>

34

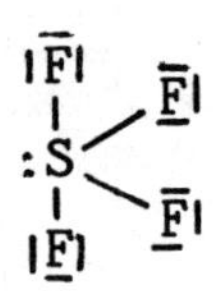

4.69 In the Lewis structure for the XeF_4 molecule, the number of lone pairs of electrons around the central Xe atom is

a. 0 b. 1 c. 2 d. 3 e. 4

4.70 The number of bonds in $S_2O_3^{2-}$ is

a. 3 b. 4 c. 5 d. 6 e. 8

Resonance

4.71 The concept of resonance describes molecular structures

a. which have several different geometrical arrangements.

b. which have no suitable single Lewis electron-dot formula.

c. which have electrons resonating.

d. which are formed from hybridized orbitals.

e. which have resonating structures.

4.72 The carbonate ion, CO_3^{2-}, is known to be planar with all the oxygen atoms equidistant from the central carbon atom. On the basis of these facts, which of the following conclusions may be drawn concerning this ion? Choose the answer that includes all the correct conclusions, and no others.

I. It can be represented by three equivalent resonance structures.

II. The dipoles associated with each C-O bond are equal in magnitude.

III. The bond orders for each C-O bond are not equivalent.

IV. The carbon atom is sp hybridized.

a. I and II b. II and III c. I and III d. I, III and IV

e. II, III and IV

84% 4.69 (c) <u># valence electrons</u>

1-Xe	:	8
4-F	:	28
		36

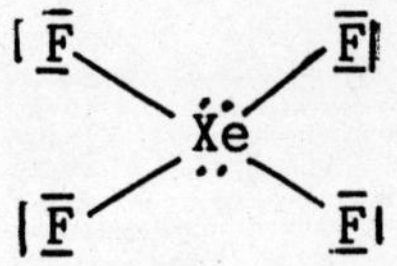

80% 4.70 (b) <u># valence electrons</u>

2-S	:	12
3-O	:	18
2e	:	2
		32

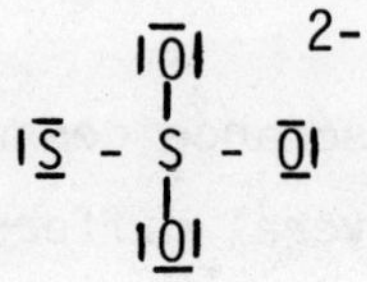

4.71 (b) The actual distribution of electrons which cannot be depicted by a single Lewis structure is an average of the various Lewis structures which can be drawn for a molecule or ion.

4.72 (a) The bond orders are all equivalent and the carbon atom is sp^2 hybridized.

4.73 Which two of the five structures shown below are the two resonance forms of N_2O?

I. :N≡N=Ö:

II. :N̈=N=Ö:

III. :N̈-N≡O:

IV :N̈-N̈-Ö:

V :N≡N-Ö̤:

a. I and II b. II and V c. III and IV d. I and III e. IV and V

4.74 Which of the following species is best described by more than one electronic structure?

a. CO_2 b. CO_3^{2-} c. SO_3^{2-} d. SO_4^{2-} e. ClO_4^-

4.75 For which of the following ions are resonance structures necessary to describe the bonding adequately?

a. NO_3^- b. ClO_3^- c. SO_4^{2-} d. SO_3^{2-} e. CN^-

Multiple Bonds

4.76 All of the following species have some multiple bonds EXCEPT

a. NO_3^- b. NO_2^- c. NO_2^+ d. SO_3^{2-} e. SO_3

4.77 Of the following, the only molecule that <u>does not</u> contain a double or triple bond is

a. CO b. CO_2 c. HCN d. P_4 e. N_2

74% 4.73 (b) Structure III and IV do not have enough valence electrons. The octet rule is violated by the central nitrogen atom in Structure I.

4.74 (b)

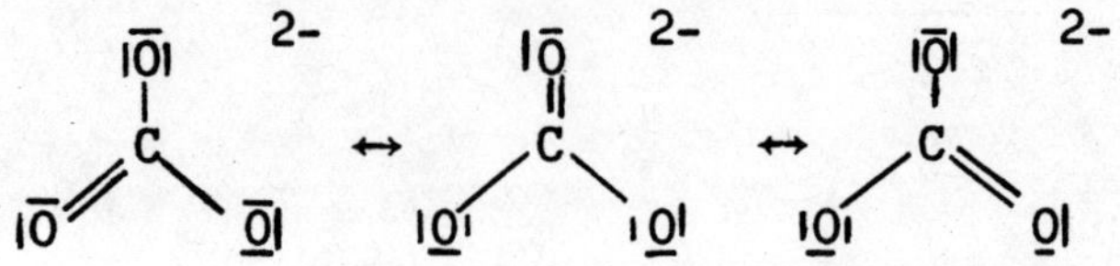

68% 4.75 (a)

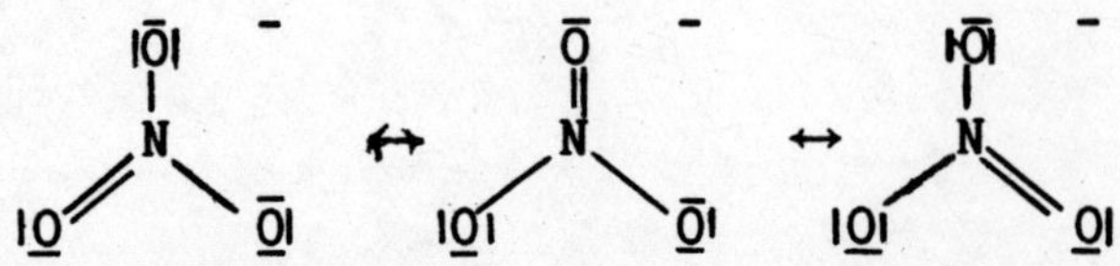

48% 4.76 (d) It is best to write the Lewis structures.

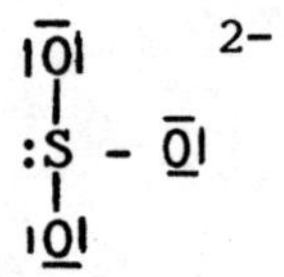

70% 4.77 (d) Students will probably have trouble drawing the Lewis structure of P_4. After studying some phosphorus chemistry, then the structure of P_4 can be obtained. The Lewis structures of the other molecules show that they have multiple bonds so one can get the answer by elimination.

:C≡O: , |$\overline{O}$=C=$\overline{O}$| , H-C≡N: , :N≡N:

4.78 Which of the following molecules contains <u>one</u> triple bond?

a. NH_3 b. CO c. H_2CCO d. H_2CCH_2 e. HNNH

4.79 All of the following molecules contain two pi-bonds EXCEPT

a. OCO b. H_2CCH_2 c. H_2CCCH_2 d. H_2CCO e. NNO

4.80 The Lewis structure for each of the following species contains a triple bond EXCEPT

a. NO^+ b. HCN c. O_2^{2+} d. CO e. N_3^-

4.81 The Lewis structure for each of the following contains a triple bond EXCEPT

a. CaC_2 b. N_2 c. CO d. CN^- e. NO^-

4.82 Which of the following species has the shortest bond distance?

a. CO b. NO^- c. O_2 d. F_2 e. Cl_2

4.83 In 1, 1, 2, 2-tetracyanoethylene, $(NC)_2CC(CN)_2$, the number of π-bonds, σ-bonds, and lone pairs of electrons in the molecule, respectively, is

	π-bonds	σ-bonds	lone pairs of electrons
a.	5	13	6
b.	6	12	6
c.	8	10	4
d.	9	9	4
e.	12	6	6

Shapes of Molecules

4.84 All of the following species are linear EXCEPT

a. OSO b. HCCH c. NNO d. OCO e. HCN

52% 4.78 (b) :C≡O:

4.79 (b) 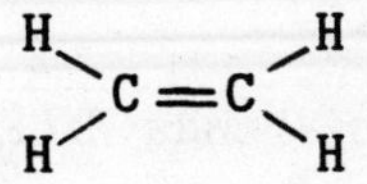 Ethene has one double bond. This consists of one pi-bond and one sigma bond.

44% 4.80 (e) <u># valence electrons</u>

3-N : 15

1e : <u>1</u>

16

$:\ddot{N}=N=\ddot{N}:^{-}$

76% 4.81 (e) C_2^{2-}, N_2, CO and CN^- all have 10 valence electrons and a triple bond. NO^- has 12 valence electrons and is isoelectronic with O_2.

64% 4.82 (a) Which species has the most multiple bonding or the highest bond order?

CO	a triple bond
NO^-	a double bond
O_2	a double bond
F_2	a single bond
Cl_2	a single bond

4.83 (d) 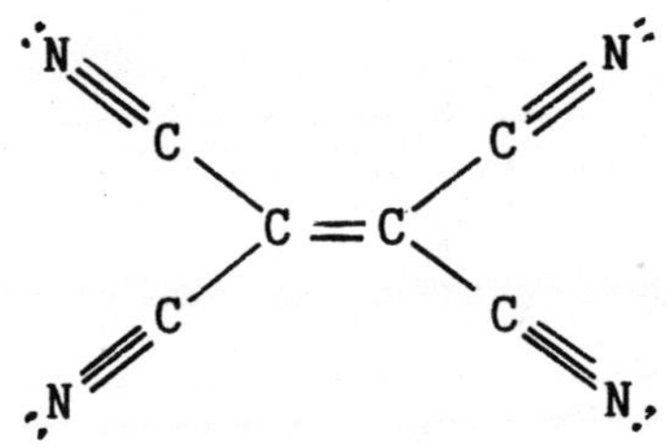

4.84 (a) 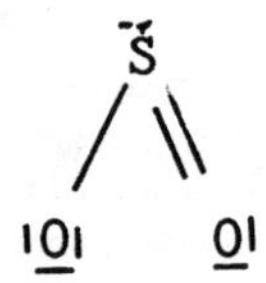 The lone pair of electrons on sulfur causes a bent molecule. In all the other molecules in this question, the central atom does not have any lone pairs of electrons (only bonding pairs).

4.85 All of the following species are bent EXCEPT

a. HOH b. HSH c. ONO^- d. ICl_2^- e. OSO

4.86 All of the following species would be expected to have a tetrahedral shape EXCEPT

a. ICl_4^- b. ClO_4^- c. NH_4^+ d. SiH_4 e. PO_4^{3-}

4.87 Which one of the following molecules has a flat triangular structure?

a. NH_3 b. SO_3 c. CCl_4 d. HCN e. AsF_3

4.88 All of the following species are planar EXCEPT

a. H_2NNH_2 b. H_2CCH_2 c. FNNF d. H_2CO e. CO_3^{2-}

4.89 Which one of the following species would not be expected to have a tetrahedral shape?

a. SiF_4 b. BeF_4^{2-} c. BF_4^- d. NF_4^+ e. XeF_4

4.90 All of the following species are planar EXCEPT

a. BF_3 b. F_2CO c. SF_4 d. XeF_4 e. F_2CCF_2

4.91 All the following species are planar EXCEPT: (Hint: First write their Lewis structures.)

a. F_2CCF_2 b. Cl_2CO c. ClO_3^- d. ICl_4^- e. SO_3

Polarity of Molecules

4.92 Of the following molecules, which would have a measurable dipole moment for the molecule?

a. 

b. CS_2 c. CCl_4 d. HCCH e. SO_2

64% 4.85 (d) 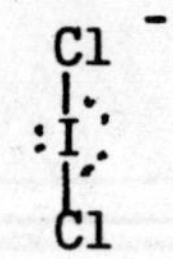 The central distribution of the lone pairs of electrons about the central atom results in the formation of a linear species.

60% 4.86 (a) 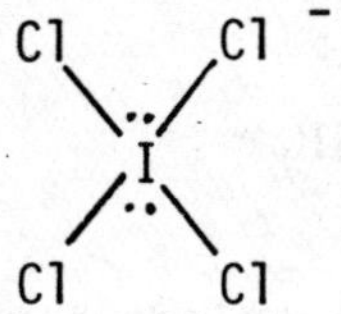 ICl_4^- is a planar ion. Repulsion is minimized by a pair of electrons above and below the plane of the ion.

56% 4.87 (b)  Repulsion is minimized by the three regions of electron density forming a trigonal plane.

40% 4.88 (a) 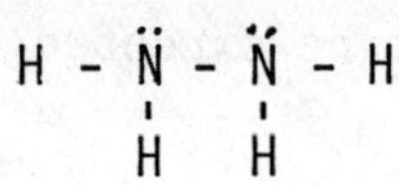 The lone pairs of electrons on nitrogen causes each nitrogen atom to be in a pyramidal configuration.

80% 4.89 (e) All the species from (a) - (d) have four pairs of electrons around the central atom. Xe in XeF_4 has 6 pairs of electrons around it so that XeF_4 is square planar.

32% 4.90 (c) BF_3 and F_2CO are trigonal planar. SF_4 has a distorted tetrahedral shape since there are four bonding pairs of electrons plus a lone pair of electrons about the central sulfur atom.

52% 4.91 (c) The central Cl atom in ClO_3^- has four regions of electron density about it. The lone pair of electrons on Cl gives the ion a pyramidal shape.

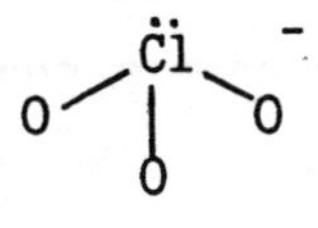

4.92 (e) The molecules in distractors (a) - (e) all have polar bonds, but because of their symmetry, the bond dipoles sum to zero. In SO_2, the sulfur has a lone pair of electrons and the molecule is nonlinear.

4.93 Of the following molecules, the only one with no permanent dipole moment is

a. water, H_2O b. acetone, CH_3COCH_3 c. carbon dioxide, CO_2
d. sulfur dioxide, SO_2 e. 1,2-dichlorobenzene, $C_6H_4Cl_2$

4.94 All of the following molecules would be polar EXCEPT

a. O_3 b. SO_3 c. H_2O d. OF_2 e. F_2CO

4.95 Of the following molecules, which would have a measurable dipole moment for the molecule?

a. HOOH b. CO_2 c. HCCH d. SF_6 e. PF_5

4.96 Which response contains all the characteristics listed that should apply to phosphine, PH_3?

I. trigonal planar
II. one unshared pair of electrons on P
III. sp^2 hybridized phosphorus atom
IV. polar molecule
V. polar bonds

a. I, IV and V b. II, III and IV c. I, II and IV
d. II, IV and V e. I, III and V

4.97 Which one of the following molecules has a non-zero dipole moment?

a. BCl_3 b. CCl_4 c. $SiCl_4$ d. SF_6 e. NCl_3

4.93 (c) $|\overline{O} = C = \overline{O}|$, CO_2 is a symmetrical linear molecule.

30% 4.94 (b) SO_3 is a trigonal planar molecule with polar bonds, but because of the symmetry, the bond dipoles sum to zero.

30% 4.95 (a) H - Ö - Ö - H   Hydrogen peroxide is nonlinear and polar.

4.96 (d) H - P̈ - H (with H below P)   Phosphine has a pyramidal shape and the phosphorus atom is hybridized sp^3.

4.97 (e) Which molecule is polar? Which molecule has a central atom with a lone pair of electrons around it?

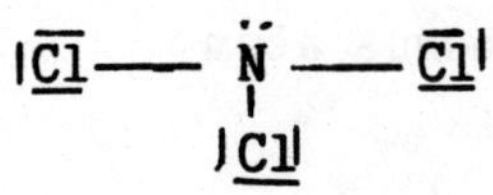

The following five sets of compounds or elements are to be used for questions 4.98 - 4.100.

a. S_8, P_4, N_2, F_2
b. $NaClO_4$, PF_3, CS_2, Fe_2O_3
c. Na_2O, CaH_2, BaO, KF
d. CH_4, H_2CCl_2, N_2, Cl_2
e. CH_3OH, H_2O, HF, PF_3

4.98 Which set contains only ionic compounds?

4.99 Which set contains only polar covalent substances?

4.100 Which set contains only nonpolar species?

Valence Bond Theory, Hybridization

4.101 What is the hybridization of the central atom in a tetrahedral molecule?

a. p^3 b. sp c. sp^2 d. sp^3 e. s^2p^2

4.102 The hybridization of the central atom in a molecule is described as sp^2. The arrangement in space of the hybrid orbitals about that atom is

a. linear b. trigonal planar c. tetrahedral
d. trigonal bipyramidal e. octahedral

4.103 The geometric arrangement of sp^3d^2 orbitals is

a. linear b. trigonal planar c. tetrahedral
d. trigonal bipyramidal e. octahedral

64% 4.98 (c) The series of compounds which in each case consists of a metal and a nonmetal.

32% 4.99 (e) The series which consists of nonmetals bonded to nonmetals and has unsymmetrical distribution of the electrons about the central atom.

4.100 (a) Like atoms bonded to each other in a symmetrical way.

4.101 (d)

4.102 (b)

4.103 (e)

4.104 When a carbon atom has sp^3 hybridization it has

a. 4 π bonds b. 3 π bonds and 1 σ bond

c. 2 π bonds and 2 σ bonds d. 1 π bond and 3 σ bonds

e. 4 σ bonds

4.105 From a consideration of the three molecules below, a compound(s) which contain(s) both sp and sp^3 hybridized carbon atoms is(are)

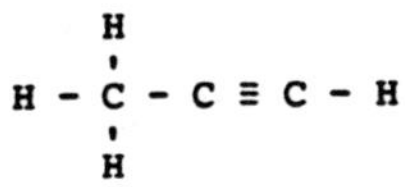

1

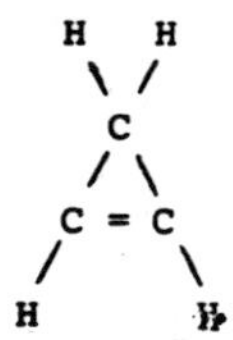

2

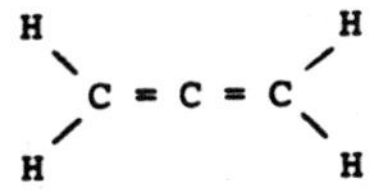

3

a. 1 only b. 2 only c. 3 only d. 1 and 2 only

e. 1 and 3 only

4.106 What hybrid orbitals of beryllium are involved in the bonding in gaseous BeF_2?

a. sp b. sp^2 c. sp^3 d. dsp^2 e. d^2sp^3

4.107 The electron pairs on the central N can be considered to be in sp^2 hybridized orbitals in all the following species EXCEPT

a. FNO b. NO_3^- c. H_2NNH_2 d. HNNH e. HONO

4.104 (e)

4.105 (a)

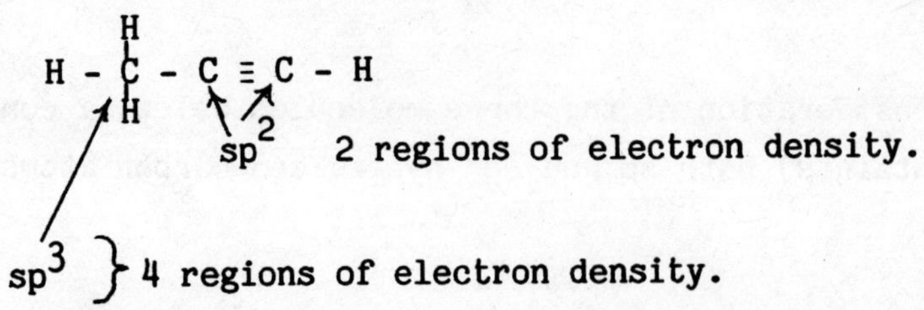

4.106 (a) F : Be : F } 2 regions of electron density about the central atom.

48% 4.107 (c)

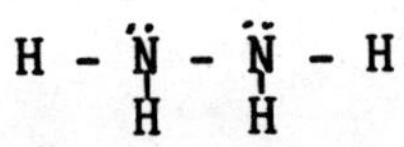

There are four regions of electron density about each nitrogen, so nitrogen atoms can be described as being sp^3 hybridized.

Questions 4.108 - 4.112 pertain to the bonding and structure of the carbonate ion.

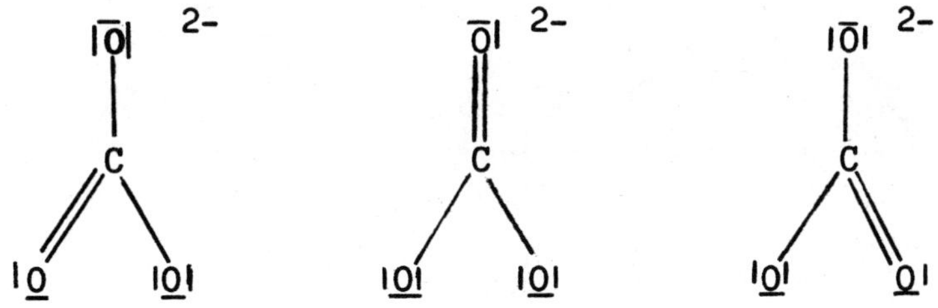

x y z

4.108 The fact that all the bonds in CO_3^{-2} are the same is accounted by a consideration of structure(s)

a. y only b. x and y c. x and z d. y and z
e. x, y and z

4.109 The expected C-O bond order is

a. 1 b. 1-1/3 c. 1-1/2 d. 1-2/3 e. 2

4.110 The hybridization of the orbitals on the carbon is described as

a. s b. sp c. sp^2 d. sp^3 e. p^3

4.111 The O-C-O bond angle(s) is

a. 90° b. 109° c. 120° d. 180° e. 180° and 90°

4.112 The shape of the carbonate ion is best described as

a. linear b. planar c. pyramidal d. tetrahedral
e. trigonal pyramidal

4.108 (e) The actual structure of CO_3^{2-} is a resonance hybrid of structures x, y and z.

4.109 (b) $\frac{\text{\# of bonds}}{3} = \frac{4}{3}$

4.110 (c) $\underline{\uparrow}\ \underline{\uparrow}\ \underline{\uparrow}\ \underline{\uparrow}$

2s 2p

sp^2

There are three regions of electron density about the central carbon atom.

4.111 (c) sp hybridization bond angles 180°

sp^2 hybridization bond angles 120°

sp^3 hybridization bond angles 109°

4.112 (b)

Molecular Orbital Theory

The energy level diagram could be useful for questions 4.113 - 4.116.

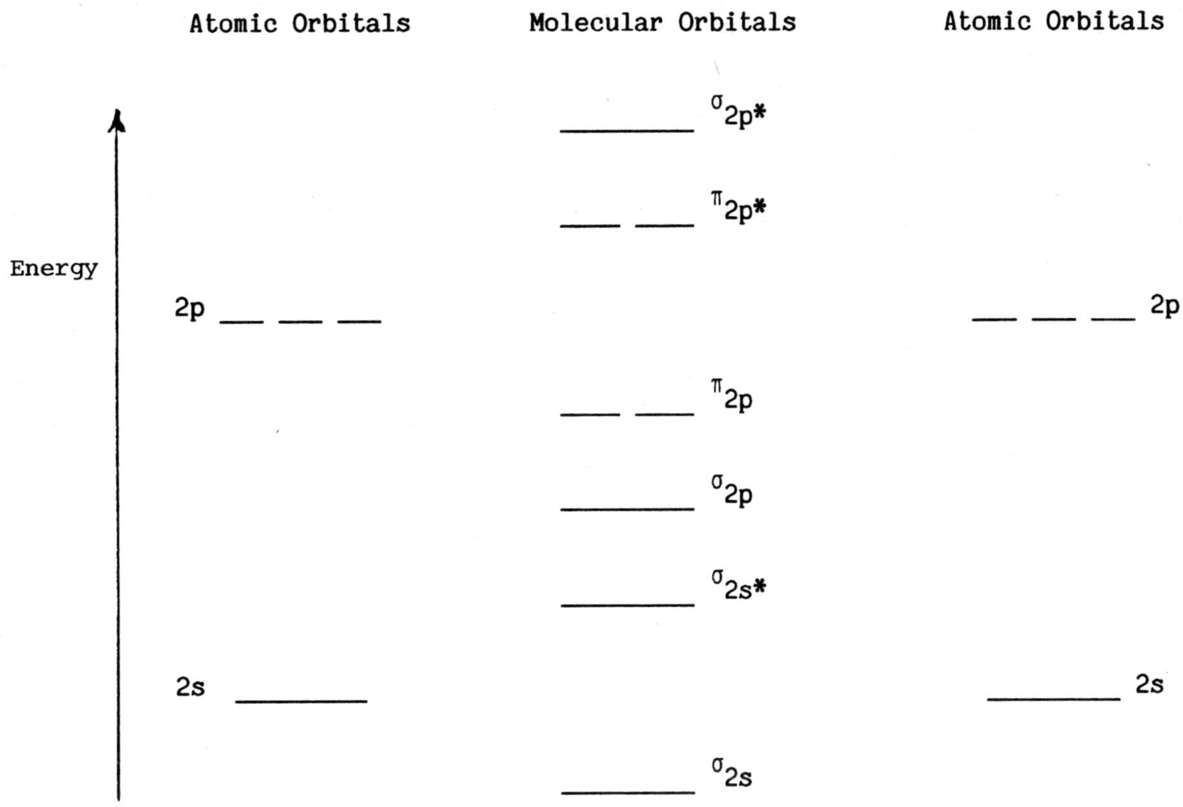

4.113 The bond order for O_2^+ is

a. 1 b. 1.5 c. 2 d. 2.5 e. 3

4.114 According to the above energy level diagram, the gaseous species NO^+ in the ground state has a bond order of ________ and is ________.

a. 2, diamagnetic b. 2, paramagnetic c. 2.5, paramagnetic

d. 3, diamagnetic e. 3, paramagnetic

4.113 (d) Bond order = $\frac{\text{\# bonding electrons} - \text{\# antibonding electrons}}{2}$

Bond order = $\frac{8 - 2}{2}$

4.114 (d) Bond order = $\frac{8 - 2}{2}$

All the electrons are paired so the ion is diamagnetic.

4.115 For which of the following endothermic processes would the energy requirement be the most?

I. $CO(g) \rightarrow C(g) + O(g)$

II. $CO^{-}(g) \rightarrow C(g) + O^{-}(g)$

III. $CO^{+}(g) \rightarrow C^{+}(g) + O(g)$

a. I b. II c. III d. II and III e. I and III

4.116 For which of the following endothermic processes would the energy requirement be the largest?

I. $NO^{-}(g) \rightarrow N(g) + O^{-}(g)$

II. $NO(g) \rightarrow N(g) + O(g)$

III. $NO^{+}(g) \rightarrow N^{+}(g) + O(g)$

a. I b. II c. III d. All are equal.

e. Need more information.

Bond Energies

4.117 The bond energy of hydrogen bromide, HBr, is the change in enthalpy for the reaction

a. $1/2\ H_2(g) + 1/2\ Br_2(l) \rightarrow HBr(g)$

b. $H_2(g) + Br_2(l) \rightarrow 2\ HBr(g)$

c. $HBr(g) \rightarrow H(g) + Br(g)$

d. $HBr(g) \rightarrow 1/2\ H_2(g) + Br(g)$

e. $HBr(g) \rightarrow 1/2\ H_2(g) + 1/2\ Br_2(g)$

4.115 (a) Which species, CO, CO^- or CO^+ has the highest bond order?

Species	Bond Order
CO	$\frac{8-2}{2} = 3$
CO^-	$\frac{8-3}{2} = 2.5$
CO^+	$\frac{7-2}{2} = 2.5$

46% 4.116 (c) Which species, NO^-, NO or NO^+ has the highest bond order?

Species	Bond Order
NO^-	$\frac{8-4}{2} = 2$
NO	$\frac{8-3}{2} = 2.5$
NO^+	$\frac{8-2}{2} = 3$

4.117 (c) The bond dissociation energy is the energy required to break a gaseous molecule into gaseous atoms.

4.118 The bond energy of carbon dioxide, CO_2, is the change in enthalpy for the reaction

a. $CO_2(g) \rightarrow C(g) + 2\ O(g)$

b. $CO_2(g) \rightarrow C(g) + O_2(g)$

c. $CO_2(g) \rightarrow C(s) + O_2(g)$

d. $CO_2(g) \rightarrow C(s) + 2\ O(g)$

e. $C(s) + O_2(g) \rightarrow CO_2(g)$

4.119 Given the reactions

$NF_3(g) \rightarrow N(g) + 3F(g), \quad \Delta H = 841\ kJ$

$FN = NF \rightarrow 2N(g) + 2F(g), \ \Delta H = 1033\ kJ$

The N=N bond energy in kJ/mol is

a. 192 b. 280 c. 473 d. 560 e. 1874

4.120 Given the reactions

$CH_4(g) \rightarrow C(g) + 4H(g) \quad \Delta H = 1656\ kJ$

$HC{\equiv}CH(g) \rightarrow 2C(g) + 2H(g) \quad \Delta H = 1648\ kJ$

The C≡C bond energy in kJ/mol is

a. 414 b. -8 c. -1664 d. 1664 e. 820

4.121 The bond energies for $H_2(g)$ and $HCl(g)$ are 435 kJ/mol and 431 kJ/mol respectively. For the reaction $1/2\ H_2(g) + 1/2\ Cl_2(g) \rightarrow HCl(g)$ the heat of reaction is -92 kJ/mol of HCl produced. The bond energy of Cl_2 in kJ/mol is

a. 88 b. 96 c. 188 d. 243 e. 335

80% 4.118 (a) See question 4.117.

48% 4.119 (c) The average N-F bond energy is 841/3 = 280 kJ.

2 N-F = 560 kJ

N=N = y

FN = NF = 1033 kJ

y = 1033 - 560 = 473 kJ

50% 4.120 (e) The average C-H bond energy is 1656/4 = 414 kJ.

2 C-H = 828 kJ

C≡C = y

HC≡CH = 1648 kJ

y = 1648 - 828 = 820 kJ

4.121 (d) ΔH_{rxn} = 1/2 B.E. (H_2) + 1/2 B.E. (Cl_2) - B.E. (HCl)

-92 = 435/2 + 1/2 B.E. (Cl_2) - 431

1/2 B.E. (Cl_2) = 121.5 kJ

B.E. (Cl_2) = 243 kJ

5

SOLUTION CALCULATIONS

Percent Composition

5.1 If 32.0 g of NaCl is dissolved in 350. g of H_2O, what is the PERCENT BY WEIGHT (by mass) of NaCl in the solution?

a. 0.0838% b. 0.0914% c. 10.7% d. 9.14% e. 8.38%

5.2 If. 0.200 mol of copper(II) sulfate pentahydrate, $CuSO_4 \cdot 5H_2O$, is dissolved in 275 g of H_2O, what is the PERCENT BY WEIGHT of the nonhydrated copper(II) sulfate, $CuSO_4$, in the solution (At. wt: Cu = 63.6, S = 32.0, H = 1.00)?

a. 9.82% b. 10.4% c. 10.9% d. 15.4% e. 16.3%

5.3 What mass of 45.0% ammonium nitrate, NH_4NO_3, <u>solution</u> contains 80.0 g of water?

a. 36 g b. 44 g c. 145 g d. 178 g e. 215 g

5.4 What is the mass of H_2SO_4 in a 50.0 cm^3 sample of sulfuric acid which has a density of 1.55 g cm^{-3} and which consists of 65.0% H_2SO_4?

a. 27.1 g b. 32.5 g c. 50.4 g d. 77.5 g e. 119 g

The following information pertains to questions 5.5 and 5.6.
76 g of NH_3 is dissolved in 194 g of water. The final volume of the solution is 3.00 x 10^2 ml.

5.5 The percentage by mass of the solute is

a. 64.7 b. 39.2 c. 28.1 d. 25.3 e. None of these.

5.6 The density of the solution in g/ml is

a. 1.39 b. 0.900 c. 0.647 d. 0.390 e. None of these.

60% 5.1 (e) $\frac{32.0}{350 + 32.0} \times 100$

32% 5.2 (a) $\# \text{ mol } H_2O = \frac{5\ H_2O}{1\ CuSO_4 \cdot 5H_2O} \times 0.200 \text{ mol } CuSO_4 \cdot 5H_2O = 1.00 \text{ mol}$

$$\text{wt } H_2O - 275 \text{ g} + 1.00 \text{ mol } H_2O \times \frac{18.0 \text{ g } H_2O}{1 \text{ mol } H_2O} = 293 \text{ g}$$

$$\text{wt } CuSO_4 = 0.200 \text{ mol } CuSO_4 \times \frac{159.6\ CuSO_4}{1 \text{ mol } CuSO_4} = 31.9 \text{ g}$$

$$\%\ CuSO_4 = \frac{31.9}{293 + 31.9} \times 100$$

52% 5.3 (c) $\frac{y}{80 + y} = 0.45$; $y = 0.45\ (80 + y)$

y = mass solute = 65 g

Mass of solution = mass of solute + mass H_2O

= 65 + 80

80% 5.4 (c) $\text{wt} = 50.0 \text{ cm}^3 \times 1.55 \text{ g/cm}^3 \times 0.65$

5.5 (c) $\%\text{ solute} = \frac{\text{mass solute}}{\text{mass solute + mass solvent}} \times 100$

$$= \frac{76}{76 + 194} \times 100$$

5.6 (b) $d = \frac{M}{V} = \frac{76 + 194 \text{ g}}{3.00 \times 10^2 \text{ mL}}$

Molarity

5.7 The molarity of a solution is defined as

a. the number of moles of solute per kilogram of solvent.

b. the number of moles of solute per liter of solution.

c. the number of gram-equivalent weights of solute per liter of solution.

d. the number of gram-equivalent weights of solute per kilogram of solution.

e. the number of moles of solute per liter of solvent.

5.8 A student needs to prepare 2.00 L of 0.100 M Na_2CO_3 (molecular mass = 106). The best procedure is to weigh out

a. 21.2 g Na_2CO_3 and add H_2O until the final homogeneous solution has a volume of 2.00 L.

b. 10.6 g Na_2CO_3 and add H_2O until the final homogeneous solution has a volume of 2.00 L.

c. 21.2 g Na_2CO_3 and add 2.00 kg of H_2O.

d. 10.6 g Na_2CO_3 and add 2.00 kg of H_2O.

e. 21.2 g Na_2CO_3 and add 2.00 L of H_2O.

5.9 The value of a solution concentration unit which may change with temperature is

a. molarity b. mole fraction c. molality

d. mole percentage e. weight percentage

5.10 Calculate the molarity of a solution that contains 56.0 g of HCl in 265 mL of solution. [Mol. Wt. HCl = 36.5 g/mole.]

a. 2.94 b. 5.79 c. 6.35 d. 7.06 e. 7.60

5.11 A 60.0 g sample of NaOH is dissolved in water and the solution is diluted to give a final volume of 3.00 liters. The MOLARITY of the final solution is (Mol. Wt. NaOH = 40.0)

a. 0.500 M b. 0.750 M c. 1.00 M d. 2.00 M e. 3.00 M

5.7 (b)

84% 5.8 (a) $0.100 \frac{mol}{L} \times 2.00\ L \times 106 \frac{g}{mol} = 21.2\ g$

78% 5.9 (a) The density of a solution is dependent upon temperature (the volume of the solution will change with temperature). All the other concentration units (b) - (d) are weight/weight units.

96% 5.10 (b) $M = \frac{\#\ moles}{Volume\ in\ liters}$

$$M = \frac{56.0\ g\ HCl \times \frac{1\ mol\ HCl}{36.5\ g\ HCl}}{0.265\ L}$$

96% 5.11 (a) $M = \frac{60.0\ g\ NaOH \times \frac{1\ mol\ NaOH}{40.0\ g\ NaOH}}{3.00\ L}$

5.12 How many g of KOH are contained in 400. mL of 0.250 M KOH solution?

a. 5.61 g b. 8.98 g c. 12.4 g d. 35.1 g e. 89.8 g

5.13 A sodium nitrate, $NaNO_3$ (M.W. = 84.99) solution is 1.52 molar. A 35.0 ml sample of this solution contains ____ moles of $NaNO_3$.

a. 4520 b. 53.2 c. 4.52 d. 1.52 e. 0.0532

5.14 A 54.2 g sample of barium hydroxide, $Ba(OH)_2$, is dissolved in water and the solution is diluted to give a final volume of 5.00 liters. The MOLARITY of the hydroxide ion, OH^- in the final solution is (M.W. $Ba(OH)_2$ = 171).

a. 0.0317 b. 0.0634 c. 0.127 d. 0.190 e. 0.634

5.15 The number of moles of sulfate ions in a 0.1 liter solution of 0.02 M $Al_2(SO_4)_3$ is

a. 0.002 mole b. 0.004 mole c. 0.006 mole d. 0.024 mole

e. 0.06 mole

The data below are for questions 5.16 - 5.18. Concentrated nitric acid is 69.0% HNO_3 by mass and it has a density of 1.41 g/ml at 20°C. M.W. of HNO_3 = 63.0 g/mole.

5.16 A liter of nitric acid weighs

a. 1000 g b. 1410 g c. 1410 g x 0.69 d. 1000 g x 0.69

e. $\frac{1410 \times 0.69}{63}$

5.17 The mass of HNO_3 in each liter of concentrated HNO_3 is

a. (1410 x 0.690)g b. (1410 - 1000)g c. 1410 g

d. (1410 x 0.690-1000)g e. 690 g

5.18 The molarity of the HNO_3 solution is

a. 6.5 M b. 11.0 M c. 12.0 M d. 15.4 M e. 22.4 M

5.12 (a) $\# \text{ mol KOH} = 0.250 \frac{\text{mol}}{\text{L}} \times 0.400 \text{ L} = 0.100$

$$\text{g KOH} = 0.100 \text{ mol} \times \frac{56.1 \text{ g KOH}}{1 \text{ mol KOH}}$$

5.13 (e) $\# \text{ mol} = M \times V_L$

$$= 1.52 \frac{\text{mol}}{\text{L}} \times 0.0350 \text{ L}$$

24% 5.14 (c) $\# \text{ mol Ba(OH)}_2 = 54.2 \text{ g Ba(OH)}_2 \times \frac{1 \text{ mol Ba(OH)}_2}{171 \text{ g Ba(OH)}_2}$

$$= 0.317 \text{ mol}$$

$$M_{OH^-} = \frac{\frac{2\ OH^-}{Ba(OH)_2} \times 0.317 \text{ mol Ba(OH)}_2}{5.00 \text{ L}}$$

50% of the students selected (b) as the answer. These students did not take into consideration the fact that there are 2 OH^- / $Ba(OH)_2$.

5.15 (c) $\# \text{ mol Al}_2(SO_4)_3 = 0.02 \text{ mol/L} \times 0.1 \text{ L} = 0.002$

$$\# \text{ mol } SO_4^{2-} = 0.002 \text{ mol Al}_2(SO_4)_3 \times \frac{3\ SO_4^{2-}}{1\ Al_2(SO_4)_3}$$

5.16 (b) 1000 mL x 1.41 g/mL = 1410 g

5.17 (a) 1410 g x 0.690 = 973 g

5.18 (d) $M = \frac{973 \text{ g} \times \frac{1 \text{ mol HNO}_3}{63.0 \text{ g HNO}_3}}{1.00 \text{ L}}$

5.19 An antacid tablet containing 0.50 g of $NaHCO_3$ is dissolved in 250 mL of water. What is the molar concentration of $NaHCO_3$ in the solution?

a. 0.0060 M b. 0.012 M c. 0.024 M d. 0.048 M e. 2.0 M

5.20 The volume of 0.200 molar Na_2CO_3 (FW = 106) solution that contains 53.0 g of Na_2CO_3 is

a. 0.200 L b. 0.400 L c. 0.500 L d. 1.60 L e. 2.50 L

5.21 What volume of 0.750 molar H_2SO_4 must be taken to obtain 0.273 MOLES of H_2SO_4?

a. 0.205 liters b. 0.364 liters c. 1.123 liters

d. 2.74 liters e. 3.00 liters

5.22 A 2.0 <u>molar</u> solution of H_3PO_4 will contain ____ g of H_3PO_4 in 0.25 liter of solution.

a. 196 b. 98 c. 49 d. 24 e. 12

5.23. 100. ml of a solution contains 15.8 g of HNO_3 (FW = 63.0). What is the molarity of nitric acid?

a. 0.158 M b. 2.50 M c. 0.250 M d. 0.00250M e. None of these.

5.24 How many moles of NaOH are in 22.0 ml of 0.150 M NaOH?

a. 0.00148 b. 0.00221 c. 0.00330 d. 1.48 e. 3.30

Dilutions

5.25 How much 0.540 M NaCl, "physiological saline," can be prepared by the dilution of 200. mL of a 3.00 M NaCl solution?

a. 911 mL b. 1110 mL c. 1240 mL d. 1510 mL e. 1910 mL

90% 5.19 (c) $M = \dfrac{0.50 \text{ g } NaHCO_3 \times \dfrac{1 \text{ mol } NaHCO_3}{84 \text{ g } NaHCO_3}}{0.25 \text{ L}} = 0.024$

88% 5.20 (e) $V = \dfrac{\# \text{ mol}}{M} = \dfrac{53.0 \text{ g } Na_2CO_3 \times \dfrac{1 \text{ mol } Na_2CO_3}{106 \text{ g } Na_2CO_3}}{0.200 \text{ mol/L}}$

85% 5.21 (b) $V_L = \dfrac{0.273 \text{ mol}}{0.750 \text{ mol/L}}$

5.22 (c) $\# \text{ mol } H_3PO_4 = 2.0 \dfrac{\text{mol}}{\text{L}} \times 0.25 \text{ L} = 0.50$

$\# \text{ g } H_3PO_4 = 0.50 \text{ mol } \dfrac{98 \text{ g } H_3PO_4}{1 \text{ mol } H_3PO_4}$

5.23 (b) $M = \dfrac{158 \text{ g } HNO_3 \times \dfrac{1 \text{ mol } HNO_3}{63.0 \text{ g } HNO_3}}{1.00 \text{ L}}$

5.24 (c) $\# \text{ mol NaOH} = 0.150 \text{ mol/L} \times 0.0220 \text{ L}$

96% 5.25 (b) $V_1 \times M_1 = V_2 \times M_2$

$V_2 = \dfrac{200.\ \text{mL} \times 3.00 \text{ M}}{0.540 \text{ M}}$

5.26 What volume of water must be added to 10 ml of 6 M HCl to make a solution which will be 0.3 M in hydrogen ion? (Assuming the volumes are additive.)

a. 10 ml b. 190 ml c. 200 ml d. 210 ml e. 500 ml

5.27 When 70. mL of 3.0 M Na_2SO_4 is added to 30. mL of 1.0 M NaCl, the resulting concentration of Na^+ is

a. 2.0 M b. 2.4 M c. 4.0 M d. 4.5 M e. 7.0 M

5.28 200. ml of 2.50 M H_2SO_4 is added to 300. ml of 4.00 M H_2SO_4. Assuming that the volumes are additive, the final concentration is

a. 1.70 M b. 2.40 M c. 3.20 M d. 3.40 M e. 4.00 M

5.29 To prepare 100. ml of 0.50 M HCl from 2.00 M HCl, what volume of acid must you use?

a. 25.0 ml b. 50.0 ml c. 100. ml d. 200. ml e. 400. ml

Calculations Involving Solutions

5.30 If 35.00 mL of a NaOH solution are required for the neutralization of a 25.00 mL sample of 0.2000 M H_2SO_4 which was diluted to 75 mL, what is the molarity of the NaOH?

a. 0.1429 M b. 0.2857 M c. 0.4285 M d. 0.8571 M
e. 1.1423 M

5.31 How many mL of 0.300 M NaOH are required to react with 500. mL of 0.170 M H_3PO_4 to yield Na_2HPO_4? (First write a balanced equation.)

a. 94.4 b. 142 c. 283 d. 567 e. 850

5.26 (b) $10 \text{ mL} \times 6 \text{ M} = y \text{ mL} \times 0.3 \text{ M}$

$y = 200 \text{ mL}$

H_2O added = 200 mL - 10 mL = 190 mL

Don't forget that we started with 10 mL and that the final volume is 200 mL.

40% 5.27 (d) $\# \text{ mmol } Na^+ = \frac{2 \text{ } Na^+}{1 \text{ } Na_2SO_4} \times 3.0 \text{ M } Na_2SO_4 \times 70. \text{ mL} + 1.0 \text{ M } Na^+ \times 30. \text{ mL} = 450$

$$M_{Na^+} = \frac{\# \text{mmol } Na^+}{\text{mL}} = \frac{450 \text{ mmol}}{100 \text{ mL}} = 4.5 \text{ M}$$

5.28 (d) $\# \text{ mmol } H_2SO_4 = 200. \text{ mL} \times 2.50 \text{ M} = 500$

$\# \text{ mmol } H_2SO_4 = 300. \text{ mL} \times 4.00 \text{ M} = 1200$

$$M = \frac{500 \text{ mmol} + 1200 \text{ mmol}}{200 \text{ mL} + 300 \text{ mL}} = 3.40 \text{ M}$$

5.29 (a) $100. \text{ mL} \times 0.50 \text{ M} = V \times 2.00 \text{ M}$

$$V = \frac{100. \times 0.50}{2.00}$$

5.30 (b) $2 \text{ NaOH} + H_2SO_4 \rightarrow Na_2SO_4 + 2H_2O$

$\# \text{ mol } OH^- = \# \text{ mol } H^+$

$$= \frac{2 \text{ NaOH}}{1 \text{ } H_2SO_4} \times 0.2000 \text{ mol/L} \times 0.02500 \text{ L}$$

$= 0.01000 \text{ mol}$

$$M_{NaOH} = \frac{0.01000 \text{ mol}}{0.03500 \text{ L}}$$

5.31 (d) $2 \text{ NaOH} + H_3PO_4 \rightarrow Na_2HPO_4 + 2H_2O$

$$\# \text{ mol NaOH} = \frac{2 \text{ NaOH}}{1 \text{ } H_3PO_4} \times 0.170 \frac{\text{mol}}{\text{L}} \times 0.500 \text{ L}$$

$= 0.170 \text{ mol}$

$$V_L = \frac{0.170 \text{ mol}}{0.300 \frac{\text{mol}}{\text{L}}} = 0.567 \text{ L}$$

5.32 A solution which is formed by combining 300 ml of 0.10 M $Ba(OH)_2$ with 200 ml of 0.15 M HCl has an OH^- concentration of

a. 10^{-7} M b. 0.03 M c. 0.06 M d. 3.3×10^{-13} M
e. 6.7×10^{-13} M

5.33 What volume of 0.500 molar nitric acid, HNO_3, is required to react with 3.70 g of calcium hydroxide, $Ca(OH)_2$, according to the following equation?

$$2HNO_3 + Ca(OH)_2 \rightarrow Ca(NO_3)_2 + 2H_2O$$

a. 25.0 mL b. 50.0 mL c. 82.0 mL d. 100.0 mL e. 200.0 mL

5.34 The titration of HCl with NaOH is represented by the equation

$$HCl + NaOH \rightarrow NaCl + H_2O$$

What volume of 0.100 M HCl is required to titrate 50.0 ml of 0.500 M NaOH?

a. 50.0 ml b. 100. ml c. 200. ml d. 250. ml e. 500. ml

5.35 What volume of 0.135 M HCl is required to completely neutralize 25.00 mL of 0.1030 M $Ba(OH)_2$?

a. 16.38 mL b. 19.07 mL c. 32.77 mL d. 38.15 mL e. 57.22 mL

5.36 How many grams of AgCl would be precipitated if an excess of $AgNO_3$ solution were added to 55.0 ml of 0.200 M KCl solution? [Mol. Wt. AgCl = 143.32 g/mol].

a. 1.10 g AgCl b. 1.58 g AgCl c. 6.42 g AgCl d. 7.80 g AgCl
e. 37.6 g AgCl

5.37 How many grams of $PbCl_2$ would be precipitated if an excess of $Pb(NO_3)_2$ solution were added to 75.0 mL of 0.200 M KCl solution? [Formula weight $PbCl_2$ = 278.1 g mol^{-1}]

a. 1.04 b. 2.09 c. 4.17 d. 8.34 e. 41.7

5.32 (c) $Ba(OH)_2 + 2HCl \rightarrow BaCl_2 + 2\ H_2O$

$$\#\ \text{mmol}\ OH^- = 0.10\ M \times 300\ mL \times \frac{2OH^-}{1\ Ba(OH)_2} = 60$$

$$\#\ \text{mmol}\ H^+ = 0.15\ M \times 200\ mL = 30$$

$$\text{excess}\ OH^- = 60\text{-}30 = 30\ \text{mmol}$$

$$M_{OH^-} = \frac{30\ \text{mmol}}{300\ mL + 200\ mL} = 0.060\ M$$

76% 5.33 (e) $$\#\ \text{mol}\ HNO_3 = \frac{2\ HNO_3}{1\ Ca(OH)_2} \times 3.70\ g\ Ca(OH)_2 \times \frac{1\ \text{mol}\ Ca(OH)_2}{74.0\ g\ Ca(OH)_2}$$

$$= 0.100\ \text{mol}$$

$$V_L(HNO_3) = \frac{0.100\ \text{mol}}{0.500\ \frac{\text{mol}}{L}} = 0.200\ L$$

5.34 (d) $V_{HCl} \times 0.100\ M = 50.0\ mL \times 0.500\ M$

$$V_{HCl} = \frac{50.0 \times 0.500}{0.100} = mL$$

5.35 (d) $\#\ \text{mmol}\ H^+ = \#\ \text{mmol}\ OH^-$

$$V_{HCl} \times 0.135\ M = 25.00\ mL \times 0.1030\ \frac{\text{mol}\ Ba(OH)_2}{L} \times \frac{2HCl}{1\ Ba(OH)_2}$$

$$V_{HCl} = 38.15\ mL$$

5.36 (b) $Ag^+ + Cl^- \rightarrow AgCl$

$$\#\ \text{mol}\ AgCl = \#\ \text{mol}\ Cl^-$$

$$= 0.200\ \text{mol/L} \times 0.0550\ L$$

$$= 0.0110\ \text{mol}$$

$$\#\ g\ AgCl = 0.0110\ \text{mol}\ AgCl \times \frac{143.32\ g\ AgCl}{1\ \text{mol}\ AgCl}$$

56% 5.37 (b) $Pb(NO_3)_2 + 2\ KCl \rightarrow PbCl_2 + 2\ KNO_3$

$$\#\ \text{mol}\ PbCl_2 = \frac{1\ PbCl_2}{2\ KCl} \times \#\ \text{mol}\ KCl$$

$$\#\ \text{mol}\ PbCl_2 = 1/2 \times 0.200\ \frac{\text{mol}}{L} \times 0.0750\ L$$

$$= 7.5 \times 10^{-3}$$

$$g\ PbCl_2 = 7.5 \times 10^{-3}\ \text{mol}\ PbCl_2 \times 278.1\ \text{g/mol}$$

5.38 $HNO_3(aq) + KOH(aq) \rightarrow KNO_3(aq) + H_2O$

When 80.00 mL of 0.200 M HNO_3 is added to 120.00 mL of 0.150 M KOH, the resulting solution is the same as a solution of

a. 0.160 M KNO_3 b. 0.0800 M KNO_3

c. 0.160 M KNO_3 and 0.0200 M HNO_3 d. 0.0160 M KNO_3 and 0.0200 M KOH

e. 0.0800 M KNO_3 and 0.0100 M KOH

5.39 $Ba^{2+}(aq) + 2Cl^-(aq) + 2Na^+(aq) + SO_4^{2-}(aq) \rightarrow BaSO_4(s) + 2Na^+(aq) + 2Cl^-(aq)$

Equal volumes of 0.1 M barium chloride solution and 0.2 M sodium sulfate solution are mixed together. The resulting concentration of the ions in solution are

a. 0.2 M Cl^-, 0.4 M Na^+, 0.1 M SO_4^{2-}

b. 0.05 M Ba^{2+}, 0.1 M Cl^-, 0.2 M Na^+, 0.1 M SO_4^{2-}

c. 0.1 M Ba^{2+}, 0.1 M Cl^-, 0.2 M Na^+, 0.1 M SO_4^{2-}

d. 0.1 M Cl^-, 0.2 M Na^+, 0.05 M SO_4^{2-}

e. 0.1 M Cl^-, 0.2 M Na^+

5.38 (e) Which reagent is the limiting reagent?

mmol HNO_3 = 80.00 mL x 0.200 M = 16.0

mmol KOH = 120.00 mL x 0.150 M = 18.0

HNO_3 is the limiting reagent.

mmol KNO_3 = 16.0

mmol OH^- (excess) = 18.0 - 16.0 = 2.0

$$M_{KNO_3} = \frac{16.0 \text{ mmol}}{80.0 \text{ mL} + 120.0 \text{ mL}} = 0.0800 \text{ M}$$

$$M_{OH^-} = \frac{2.0 \text{ mmol}}{80.0 \text{ mL} + 120.0 \text{ mL}} = 0.0100 \text{ M}$$

Don't forget the dilution of the products.

5.39 (d)

Ba^{2+}(a) + 2 Cl^-(aq) + 2 Na^+(aq) + SO_4^{2-}(aq) → $BaSO_4$(s) + 2 Na^+(aq) + 2 Cl^-(aq)

This is an involved problem. We must take into consideration the idea of limiting reagent and remember that there is a dilution effect. The original Cl^- concentration is 0.2 M but a two fold dilution gives 0.1 M. Likewise Na^+ originally is 0.4 M and after dilution it is 0.2 M. Ba^{2+} is the limiting reagent so there is excess SO_4^{2-}, so the only answer that qualifies is (d). Half the SO_4^{2-} is precipitated and the half that remains is diluted so that its concentration is 0.05 M.

PERIODIC RELATIONSHIPS AND CHEMICAL REACTIONS

Periodicity

6.1 All of the following statements concerning the Periodic Table are correct EXCEPT

a. Mendeleev and Meyer received most of the credit for the early development of the Periodic Table.
b. the elements in a given group or family have similar chemical properties.
c. the element at the end of each period has its outermost energy level completely filled with electrons.
d. the elements are all arranged in order of increasing atomic weights.
e. the chemical characteristics of the elements are periodic functions of their atomic numbers.

6.2 Which one of the following properties of an element is not a periodic function of the atomic number?

a. atomic volume b. atomic radii c. atomic weight
d. ionization energy e. density

6.3 All of the following are periodic functions of the atomic number EXCEPT

a. density b. molecular formula c. nuclear mass number
d. atomic volume e. boiling point

6.4 Which one of the following would be the best way for classifying the elements into groups in the Periodic Table?

a. Similarities in the first ionization energies.
b. Similarities in the formulas of the elements with oxygen.
c. Similarities in the atomic masses.
d. Measure their atomic masses.
e. Similarities in their boiling points.

6.1 (d) The elements are arranged in order of increasing atomic numbers. Some elements are heavier than the preceding element. For example, argon is heavier than potassium. What are some other examples?

76% 6.2 (c)

68% 6.3 (c) The mass number increases with increasing atomic number so there is no periodic function.

40% 6.4 (b) Group 1 M_2O

Group 2 MO

Group 3 M_2O_3

Group 4 MO_2

Periodic Relationships or Atomic Structure

6.5 Which factor parallels increasing ionization energy?

a. Increasing nonmetallic properties.

b. Increasing tendency to be oxidized.

c. Increasing atomic radii.

d. Decreasing nuclear charge.

e. Decreasing electron affinity.

6.6 Where on the Periodic Chart will the most nonmetallic elements be found?

a. Top of Group VII A b. Top of Group I A c. Top of Group I B

d. Bottom of Group VI A e. Botton of Group II A

6.7 Sodium and potassium have similar chemical and physical properties. This is best explained by the fact that both elements

a. are active metals.

b. are in period 1 of the Periodic Table.

c. have the same outer electronic configuration.

d. have low relative atomic masses.

e. have relatively low first ionization energies.

6.8 Which one of the following properties of elements is best used as a basis for classifying them into groups in the Periodic Table?

a. Electronic configurations b. Mass number

c. Relative atomic mass d. Atomic sizes e. Ionic sizes

6.5 (a) Nonmetals have small sizes, high electronegativities and high ionization energies.

6.6 (a)

80% 6.7 (c) The alkali metals are Group I not Period 1. They have the valence electronic configuration ns^1 and they all form plus one ions.

80% 6.8 (a)

Family of Elements	Valence Electronic Configuration	Positive Ions
Alkali	ns^1	M^+
Alkaline Earth	ns^2	M^{2+}
III	ns^2np^1	M^{3+}
IV	ns^2np^2	M^{4+}

Classify each of the elements into one of the four types listed below:

6.9 Element No. 20

6.10 Element No. 26

6.11 Element No. 36

6.12 Element No. 15

a. An alkaline earth metal.
b. A transition metal.
c. A non-metal.
d. A noble gas.

6.13 In the lanthanide elements, the orbitals which are partly filled with electrons are

a. 5s and 4d b. 5d and 4f c. 6s and 5d d. 6p and 5f
e. 6p and 6s

6.14 Which formula represents the most probable silicon-chlorine compound?

a. $SiCl$ b. $SiCl_2$ c. $SiCl_3$ d. $SiCl_4$ e. $SiCl_6$

6.15 Tellurium is an element in the same group as sulfur. Which of the following would be the formula of telluric acid?

a. H_2Te b. H_2TeO_2 c. H_2TeO_3 d. $H_2Te_2O_5$ e. H_2TeO_4

6.16 Which one of the following elements is most likely to form a compound of the formula H_2XO_4 where X is

a. P b. As c. Cl d. Se e. Br

6.17 Which one of the following elements is most likely to form a compound of the formula H_3XO_4 where X is

a. Cl b. S c. Se d. Br e. As

6.9 (a)

6.10 (b) Transition elements are those elements which the inner d energy level is being filled.

6.11 (d)

6. 12 (c)

6.13 (b) Some of the lanthanides have 5d electrons while all of them are filling the 4f orbitals.

6.14 (d) Silicon is in Family IV so we would expect it to have a +4 oxidation state. We could also consider that silicon would form a covalent compound and it could have four half-filled orbitals.

6.15 (e) Sulfur forms sulfuric acid, H_2SO_4, so telluric acid would be H_2TeO_4.

78% 6.16 (d) X in H_2XO_4 has a +6 oxidation state. The only element listed that has a +6 oxidation state is selenium which is from Family VI.

6.17 (e) X in H_3XO_4 has a +5 oxidation state so we would select an element from Family V, As.

6.18 What formula would be expected for a binary compound between magnesium and phosphorus?

a. MgP b. MgP_2 c. Mg_3P_2 d. Mg_2P_3 e. Mg_5P_3

6.19 Which of the following elements do you expect $_{30}Zn$ to resemble most closely in its properties?

a. $_{20}Ca$ b. $_{31}Ga$ c. $_{21}Sc$ d. $_{48}Cd$ e. $_{29}Cu$

6.20 Which pair of species has most nearly identical properties?

a. $^{1}_{1}H$ and $^{1}_{1}H^+$ b. $^{19}_{9}F$ and $^{19}_{9}F^-$ c. $^{7}_{3}Li$ and $^{23}_{11}Na$ d. $^{7}_{3}Li$ and $^{9}_{4}Be$

e. $^{12}_{6}C$ and $^{13}_{6}C$

6.21 Which pair of species has most nearly the identical properties? (Be careful - note charges.)

a. $^{1}_{1}H$ and $^{1}_{1}H^+$ b. $^{14}_{6}C$ and $^{14}_{7}N$ c. $^{7}_{3}Li$ and $^{9}_{4}Be$ d. $^{23}_{11}Na$ and $^{39}_{19}K$

e. $^{7}_{3}Li^+$ and $^{4}_{2}He$

6.22 For which of the following pairs of species is the difference in size the greatest?

a. P and Cl b. Li and B c. Li^+ and F^- d. Li^+ and Cl^-

e. O^{2-} and F^-

6.23 Of the following pairs of ions, the two which are closest in size are

a. K^+, Rb^+ b. Rb^+, Cs^+ c. Br^-, I^- d. Ti^{4+}, Zr^{4+} e. Zr^{4+}, Hf^{4+}

6.18 (c) The binary compound is an ionic compound. So the ions would be Mg^{2+} and P^{3-}.

6.19 (d) Both Zn and Cd belong to the same family, so they have the same valence electronic configuration.

40% 6.20 (e) Isotopes have nearly the same properties, as the properties are dependent upon the electronic configurations. The only difference in isotopes is the number of neutrons.

40% 6.21 (d) Both Na and K are in the same chemical family (both are alkali metals). Both are very reactive, soft metals and both form plus 1 ions.

6.22 (d) In options (a), (b) and (e) the species have fairly similar sizes because the element are close to each in the Periodic Table. Positive ions are much smaller than the parent atom and the negative ions are larger than the parent atom. Since Cl^- is larger than F^-, the logical choice is Li^+ and Cl^-.

6.23 (e) Due to the lanthanide contraction, Zr^{4+} and HF^{4+} are the same size.

Writing Net Ionic Equations

6.24 The net ionic equation for neutralization of benzoic acid by a base is

$$C_6H_5COOH(aq) + OH^-(aq) \rightarrow C_6H_5COO^-(aq) + H_2O(\ell)$$

Therefore,

a. benzoic acid is a strong acid.

b. the base is a weak base.

c. the $C_6H_5COO^-$ ion is a spectator ion.

d. the ion, K^+, could be a spectator ion.

e. the base in the titration could be NH_3.

6.25 When solutions of silver nitrate and hydrochloric acid are mixed, the spectator ions in the resulting reaction are

a. only H^+ b. only Ag^+ c. H^+ and NO_3^- d. only Ag^+ and Cl^-

e. Ag^+, H^+, NO_3^- and Cl^-

6.26 The net ionic equation for the reaction of hydrobromic acid with potassium hydroxide is

a. $HBrO_3(aq) + OH^-(aq) \rightarrow BrO_3^-(aq) + H_2O(\ell)$

b. $HBr(aq) + OH^-(aq) \rightarrow Br^-(aq) + H_2O(\ell)$

c. $H^+(aq) + OH^-(aq) \rightarrow H_2O(\ell)$

d. $HBrO_3(aq) + KOH(aq) \rightarrow KBrO_3(aq) + H_2O(\ell)$

e. $HBr(aq) + KOH(aq) \rightarrow KBr(aq) + H_2O(\ell)$

6.24 (d) The equation indicates that benzoic has been neutralized by a strong base so the base could be NaOH, KOH, or ...

6.25 (c) $Ag^+(aq) + NO_3^-(aq) + H^+(aq) + Cl^-(aq) \rightarrow AgCl(s) + H^+(aq) + NO_3^-(aq)$

6.26 (c) Hydrobromic acid is a strong acid and potassium hydroxide is a strong base.

$H^+(aq) + Br^-(aq) + K^+(aq) + OH^-(aq) \rightarrow H_2O(\ell) + K^+(aq) + Br^-(aq)$

Therefore, K^+ and Br^- are spectator ions.

6.27 The net ionic equation for the reaction which takes place upon the addition of a sodium hydroxide solution to an acetic acid solution is

a. $H_3O^+(aq) + OH^-(aq) \rightarrow 2H_2O(\ell)$

b. $Na^+(aq) + CH_3COO^-(aq) \rightarrow NaCH_3COO(aq)$

c. $CH_3COO^-(aq) + OH^-(aq) \rightarrow CH_3COOH(aq) + O^{2-}(aq)$

d. $CH_3COOH(aq) + H_2O \rightarrow H_3O^+(aq) + CH_3COO^-(aq)$

e. $CH_3COOH(aq) + OH^-(aq) \rightarrow CH_3COO^-(aq) + H_2O$

6.28 The net ionic equation for the reaction of the weak acid, phosphoric acid, with a dilute solution of calcium hydroxide to form a precipitate is

a. $6H^+(aq) + 2PO_4^{3-}(aq) + 3Ca^{2+}(aq) + 6OH^-(aq) \rightarrow Ca_3(PO_4)_2(s) + 6H_2O(\ell)$

b. $2H^+(aq) + 4H_2PO_4^-(aq) + 3Ca^{2+}(aq) + 6OH^-(aq) \rightarrow Ca_3(PO_4)_2(s) + 6H_2O(\ell)$

c. $2H_3PO_4(aq) + 3Ca(OH)_2(aq) \rightarrow Ca_3(PO_4)_2(s) + 6H_2O(\ell)$

d. $2H_3PO_4(aq) + 3Ca^{2+}(aq) + 6OH^-(aq) \rightarrow 3Ca^{2+}(aq) + 2PO_4^{3-}(aq) + 6H_2O(\ell)$

e. $2H_3PO_4(aq) + 3Ca^{2+}(aq) + 6OH^-(aq) \rightarrow Ca_3(PO_4)_2(s) + 6H_2O(\ell)$

90% 6.27 (e) Acetic acid is a weak acid so it is written in molecular form. Sodium hydroxide is a strong base so it is written as ions.

48% 6.28 (e) Weak acids are written in molecular form and strong bases are written as ions. Precipitates are shown in molecular form.

6.29 The net ionic equation for the acid-base reaction between nitrous acid and lithium hydroxide is

a. $HNO_2(aq) + LiOH(aq) \rightarrow LiNO_2(aq) + H_2O(\ell)$

b. $H^+(aq) + OH^-(aq) \rightarrow H_2O(\ell)$

c. $HNO_2(aq) + OH^-(aq) \rightarrow NO_2^-(aq) + H_2O(\ell)$

d. $H^+(aq) + NO_2^-(aq) + Li^+(aq) + OH^-(aq) \rightarrow LiNO_2(aq) + H_2O(\ell)$

e. $H^+(aq) + NO_2^-(aq) + Li^+(aq) + OH^-(aq) \rightarrow Li^+(aq) + NO_2^-(aq) + H_2O(\ell)$

6.30 The net ionic equation for the reaction of a solution of H_2SO_3 with a solution of $Ba(OH)_2$ is

a. $H^+(aq) + OH^-(aq) \rightarrow H_2O(aq)$

b. $2H^+(aq) + 2OH^-(aq) \rightarrow 2H_2O(\ell)$

c. $H_2SO_3(aq) + Ba(OH)_2(aq) \rightarrow BaSO_3 + 2H_2O$

d. $2H^+ + SO_3^{2-} + Ba^{2+} + 2OH^- \rightarrow BaSO_3 + 2H_2O$

e. $H_2SO_3 + Ba^{2+} + 2OH^- \rightarrow BaSO_3 + 2H_2O$

6.31 If ten drops of 0.10 M H_2SO_4 are added to a mL of 0.20 M $Ba(NO_3)_2$, a white precipitate forms. Which of the following formulas should appear in the <u>net ionic</u> equation for this reaction?

a. $BaSO_4$ b. HNO_3 c. $HNO_2 + H_2O$ d. $HNO_3 + Ba^{2+}$ e. $HNO_3 + SO_4^{2-}$

6.29 (c) Nitrous acid is a weak acid and lithium hydroxide is a strong base.

6.30 (e) H_2SO_3 is a weak acid so it is written as a molecule.
$Ba(OH)_2$ is a strong soluble base so it is written as ions.
$BaSO_3$ is insoluble and it is written in molecular form.

6.31 (a) $Ba^{2+}(aq) + SO_4^{2-}(aq) \rightarrow BaSO_4(s)$

Oxidation Numbers

6.32 All of the following species can have the noted charges or oxidation numbers EXCEPT

a. F^- and Mg^{2+} b. Cl(VII) and Mn(VII) c. Mn^{2+} and Cu^+

d. Al^{3+} and Sc^{2+} e. S(VI) and S^{2-}

6.33 The average oxidation number of S in $S_4O_6^{\ 2-}$ is

a. 2 b. 2.5 c. 4 d. 4.5 e. 6

6.34 Of the following compounds, which ones contain nitrogen in a 3+ oxidation state?

I. O=N - Cl II. H - O - N=O III. F_3NO IV. NH_4NO_3

a. I and II only b. II and III only c. III and IV only

d. I anv IV only e. II, III and IV only

6.35 In which of the following species does the underlined element have an oxidation number of +4?

a. $\underline{V}O_2^{\ +}$ b. $\underline{P}H_4^{\ +}$ c. $\underline{V}O^{2+}$ d. $\underline{N}O_2^{\ -}$ e/ $\underline{N}O_2^{\ +}$

6.32 (d) Sc is in family III and it has only a +3 oxidation state.

6.33 (b) $4\,y + 6(-2) = -2$

$4\,y = 12-2 = 10$

$y = 10/4$

85% 6.34 (a) ONCl , $y + (-2) + (-1) = 0$

$y = +3$

HONO , $y + 2(-2) + (1) = 0$

$y = +3$

94% 6.35 (c) VO^{2+} , $y + (-2) = +2$

$y = +4$

Classification of Chemical Reactions - Acid-Base Reaction

6.36 Which of the following reactions involve neither oxidation nor reduction?

a. $2H_2O_2 \rightarrow 2H_2O$

b. $2Al + Fe_2O_3 \rightarrow Al_2O_3 + 2Fe$

c. $NH_3 + HCl \rightarrow NH_4Cl$

d. $NH_4NO_3 \rightarrow N_2O + 2H_2O$

e. All of the reactions involve oxidation and reduction.

6.37 Consider the reactions represented by the following net ionic equations:

1. $Mg(s) + 2H_3O^+(aq) \rightarrow Mg^{2+}(aq) + H_2(g) + 2H_2O(\ell)$

2. $CO_3^{2-}(aq) + H_3O^+(aq) \rightarrow HCO_3^-(aq) + H_2O(\ell)$

3. $H_3O^+(aq) + OH^-(aq) \rightarrow 2H_2O(\ell)$

Which of the reactions are characterized as acid-base?

a. 3 only b. 1 and 2 c. 1 and 3 d. 2 and 3 e. 1, 2, and 3

6.38 Consider the reactions represented by the following net ionic equations:

1. $2Na(s) + 2H_3O^+(aq) \rightarrow 2Na^+(aq) + H_2(g) + 2H_2O(\ell)$

2. $O^{2-}(aq) + H_2O(aq) \rightarrow 2OH^-(aq)$

3. $Cu^{2+}(aq) + H_2S(g) \rightarrow CuS(s) + 2H^+(aq)$

The equation(s) which would be classified as acid-base reaction(s) is(are)

a. 1 only b. 2 only c. 1 and 2 only d. 1 and 3 only

e. 1, 2 and 3

6.36 (c) Reactions which involve elements undergoing change in oxidation number are oxidation-reduction reactions. Reaction (c) is an acid-base reaction (a transfer of a proton).

40% 6.37 (d) Reaction 1 is classified as an oxidation reaction, since there is a change in oxidation number.

36% 6.38 (b) Reaction 1 is an oxidation-reduction reaction.
Reaction 2 is an acid-base reaction because there is a proton transfer.
Reaction 3 is a precipitation reaction.

6.39 Which of the following reactions involves neither oxidation nor reduction?

a. $Cl_2 + H_2O \rightarrow HCl + HOCl$

b. $2Na + 2H_2O \rightarrow 2NaOH + H_2$

c. $3NO_2 + H_2O \rightarrow 2HNO_3 + NO$

d. $POCl_3 + 3H_2O \rightarrow H_3PO_4 + 3HCl$

e. $2H_2O \rightarrow 2H_2 + O_2$

6.40 Which of the following reactions involves neither oxidation nor reduction?

a. $N_2 + 3H_2 \rightarrow 2NH_3$

b. $2CrO_4^{2-} + 2H^+ \rightarrow Cr_2O_7^{2-} + H_2O$

c. $C_2H_4 + H_2 \rightarrow C_2H_6$

d. $Cu + 2Ag^+ \rightarrow Cu^{2+} + 2Ag$

e. $NH_4NO_2 \rightarrow N_2 + 2H_2O$

6.41 Which of the following reactions or half reactions involves neither oxidation nor reduction?

a. $CaO + SiO_2 \rightarrow CaSiO_3$

b. $HO_2^- + H_2O + 2e \rightarrow 3OH^-$

c. $2H_2O_2 \rightarrow 2H_2O + O_2$

d. $MnO_4^{2-} \rightarrow MnO_4^- + e$

e. $H_2 + F_2 \rightarrow 2HF$

Oxidation-Reduction Reactions

6.42 Which one of the following is an example of an oxidation-reduction reaction?

a. $NH_3(aq) + CH_3COOH(aq) \rightarrow NH_4^+(aq) + CH_3COO^-(aq)$

b. $H_2S(aq) + 2Ag^+(aq) + 2NO_3^-(aq) \rightarrow Ag_2S(s) + 2H^+(aq) + 2NO_3^-(aq)$

c. $Cl_2(aq) + H_2O \rightarrow HOCl(aq) + H^+(aq) + Cl^-(aq)$

d. $BaO_2 + H_2SO_4 \rightarrow BaSO_4 + H_2O_2$

e. $Cu(NH_3)_4^{2+} + 4CN^- \rightarrow Cu(CN)_4^{2-} + 4NH_3$

6.39 (d) Reaction (d) is a hydrolysis reaction (an acid-base reaction). There is no change in oxidation number for any of the atoms.

6.40 (b) Look for a reaction in which there is no change in oxidation numbers. All the distractors have an element as a reactant so these reactions should have a change in oxidation number.

6.41 (a) Ca, O and Si have not undergone a change in oxidation number.

52% 6.42 (c) Chlorine is both oxidized (an increase in oxidation number) and reduced (a decrease in oxidation number).

6.43 Which one of the following equations is an example of an oxidation-reduction reaction?

a. $K_2Cr_2O_7 + K_2SO_4 + H_2O \rightarrow 2K_2CrO_4 + H_2SO_4$
b. $CaCO_3 + 2HCl \rightarrow CaCl_2 + H_2O + CO_2$
c. $Mg_3N_2 + 6H_2O \rightarrow 3Mg(OH)_2 + 2NH_3$
d. $NH_4NO_2 \rightarrow N_2 + 2H_2O$
e. $POCl_3 + 3H_2O \rightarrow H_3PO_4 + 3HCl$

6.44 Which one of the following is an example of an oxidation-reduction reaction?

a. $2K_2CrO_4 + H_2SO_4 \rightarrow K_2Cr_2O_7 + K_2SO_4 + H_2O$
b. $BaCO_3 + 2HCl \rightarrow BaCl_2 + CO_2 + H_2O$
c. $N_2O_3 + H_2O \rightarrow 2HNO_2$
d. $Cu + 2H_2SO_4 \rightarrow CuSO_4 + 2H_2O + SO_2$
e. $2NH_3 + 2H_2SO_4 \rightarrow (NH_4)_2SO_4$

6.45 Which one of the following is an example of an oxidation-reduction reaction?

a. $SO_3(g) + H_2O(\ell) \rightarrow H_2SO_4(aq)$
b. $P_4O_{10}(s) + 6Na_2O(s) \rightarrow 4Na_3PO_4(s)$
c. $I_2(aq) + H_2O(\ell) \rightarrow HOI(aq) + HI(aq)$
d. $CaCO_3(s) \rightarrow CaO(s) + CO_2(g)$
e. $Al_2O_3(s) + 3H_2O(aq) + 2OH^-(aq) \rightarrow 2Al(OH)_4^-(aq)$

6.46 All of the following reactions could be classified as acid base reactions EXCEPT

a. $CaO + SiO_2 \rightarrow CaSiO_3$
b. $BF_3 + NH_3 \rightarrow F_3BNH_3$
c. $PO_4^{3-} + H_2O \rightarrow HPO_4^{2-} + OH^-$
d. $Na_2O + SO_3 \rightarrow Na_2SO_4$
e. $CH_3CHO + H_2 \rightarrow CH_3CH_2OH$

55% 6.43 (d) N in NH_4^+ is oxidized and N in NO_2^- is reduced.

76% 6.44 (d) Both copper and sulfure have undergone a change in oxidation number. All the other reactions in this question can be classified as acid-base reactions.

80% 6.45 (c) Only iodine has a change in oxidation number.

6.46 (e) Acetaldehyde has been reduced by hydrogen to ethanol.

6.47 All of the following reactions are examples of acid-base reactions EXCEPT

a. $Al(OH)_3(s) + 3H^+(aq) \rightarrow Al^{3+}(aq) + 3H_2O$

b. $Al(OH)_3(s) + OH^-(aq) \rightarrow Al(OH)_4^-(aq)$

c. $Al_2O_3(s) + 6H_2O \rightarrow 2Al(OH)_3 + 6H^+(aq)$

d. $2KO_2(s) + 2H_2O(\ell) \rightarrow 2KOH(s) + H_2O_2(\ell) + O_2(g)$

e. $POCl_3(\ell) + 3H_2O \rightarrow H_3PO_4(aq) + 3H^+(aq) + 3Cl^-(aq)$

6.48 Besides its ability to function as an acid, sulfuric acid is able to act as an oxidizing agent. Which one of the following equations illustrates this ability?

a. $CaF_2(s) + H_2SO_4(aq) \rightarrow CaSO_4(aq) + 2HF(aq)$

b. $2NH_3(g) + H_2SO_4(aq) \rightarrow 2NH_4^+(aq) + SO_4^{2-}(aq)$

c. $2H^+ + 2NaI(s) + H_2SO_4(aq) \rightarrow I_2(s) + 2H_2O + SO_2(g) + 2Na^+$

d. $C_{12}H_{22}O_{11}(s) \xrightarrow{H_2SO_4} 12C(s) + 11\ H_2O(\ell)$

e. $NaCl(s) + H_2SO_4(aq) \rightarrow NaHSO_4(aq) + HCl(aq)$

Precipitation Reactions

6.49 Which of the following ionic compounds is insoluble in water?

a. NH_4Cl b. Li_2CO_3 c. $PbSO_4$ d. KI e. Na_2S

6.50 All of the following compounds are soluble in water EXCEPT

a. NaCl b. $ZnCl_2$ c. $FeCl_2$ d. $BaCl_2$ e. Hg_2Cl_2

6.51 All of the following compounds are insoluble in water EXCEPT

a. CuS b. Hg_2Cl_2 c. $(NH_4)_2S$ d. $Zn(OH)_2$ e. $BaSO_4$

6.52 Choose the compound that is the most soluble in water.

a. HgS b. CuS c. Ag_2S d. FeS e. K_2S

48% 6.47 (d) Oxygen in KO_2 has undergone both oxidation and reduction.

75% 6.48 (c) H_2SO_4 has oxidized I^- to I_2 and while it has been reduced to SO_2.

6.49 (c) Students should know some solubility rules. All ammonium salts and alkali metal salts are soluble.

6.50 (e) All common chlorides are soluble except AgCl, $PbCl_2$ and Hg_2Cl_2.

6.51 (c) All ammonium salts are soluble.

6.52 (e) Most sulfides are insoluble. All potassiujm salts (alkali metal salts) are soluble.

6.53 When a solution of sodium chloride and a solution of ammonium nitrate are mixed,

a. one new salt is formed.

b. two new salts are formed.

c. a precipitate is formed.

d. the solution becomes acidic.

e. no reaction occurs.

6.54 A precipitate would be expected when an aqueous solution of cesium bromide is added to an aqueous solution of

a. potassium carbonate b. calcium chloride c. iron(II) sulfate

d. silver nitrate e. ammonium phosphate

6.55 Solutions of the following solutes were mixed. A precipitate would be formed by

a. $NaNO_3$ and $CuSO_4$ b. NH_4Cl and $CuSO_4$ c. NH_4Cl and $Pb(NO_3)_2$

d. NH_3 and CH_3COOH e. $K_2Cr_2O_7$ and H_2SO_4

6.56 Addition of dilute hydrochloric acid would clearly distinguish between solutions of

a. barium nitrate and sodium sulfate.

b. lead nitrate and silver nitrate.

c. mercury(I) nitrate and silver nitrate.

d. lead nitrate and barium nitrate.

e. calcium nitrate and barium nitrate.

6.57 Solutions of the following are mixed. In which case would a precipitate be least likely to form?

a. $Ca(NO_3)_2$ and Na_2CO_3 b. NH_4Cl and $MgSO_4$ c. $AgNO_3$ and NaCl

d. $AlCl_3$ and KOH e. Na_2S and $CuCl_2$

6.58 Which of the following reactions or half-reaction involves neither oxidation nor reduction?

a. $H_2 + Br_2 \rightarrow 2HBr$

b. $Br_2 + 2OH^- \rightarrow Br^- + OBr^- + H_2O$

c. $Fe^{2+} + SO_4^{2-} + H_2S \rightarrow FeS + H^+ + HSO_4^-$

d. $Fe + Cu^{2+} \rightarrow Fe^{2+} + Cu$

e. $Pb^{2+} + 2e \rightarrow Pb$

6.53 (e) No reaction occurs because all sodium salts and all ammonium salts are soluble.

52% 6.54 (d) AgCl, AgBr, and AgI salts are all insoluble.

60% 6.55 (c) $PbCl_2$, $PbBr_2$ and PbI_2 salts are all insoluble.

44% 6.56 (d) $PbCl_2$ is insoluble whereas $BaCl_2$ is soluble. All chlorides are soluble except AgCl, Hg_2Cl_2 and $PbCl_2$.

64% 6.57 (b) All ammonium salts are soluble and all chlorides are soluble except AgCl, $PbCl_2$ and Hg_2Cl_2.

75% 6.58 (c) This is a precipitation reaction. FeS is insoluble.

7

GASES

Temperature and Pressure

7.1 Liquid nitrogen has a boiling point at -196°C. This corresponds to

a. -385°F b. -353°F c. 77 K d. 153 K e. 469 K

7.2 The melting point of mercury is 235 K. This corresponds to

a. -38°C b. -20°C c. 20°C d. 38°C e. 508°C

7.3 The melting point of nitrogen is 63 K. This corresponds to

a. 63°C b. -63°C c. -210°C d. 483°C e. -336°C

7.4 At constant temperature, what is the pressure of the gas that is trapped in the end of the J-tube when h equals 45 mm?

a. 0 torr b. 45 torr c. 715 torr

d. 760 torr e. 805 torr

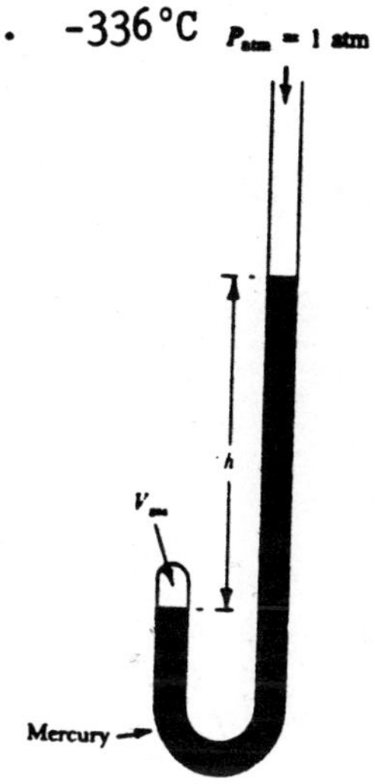

7.5 At constant temperature, the pressure of the gas that is trapped in the end of the J-tube is 760 torr. If h is equal to 50 torr, what is the pressure that is due to the atmosphere?

a. 50 torr b. 710 torr c. 760 torr d. 810 torr

e. Cannot tell because the temperature is not given.

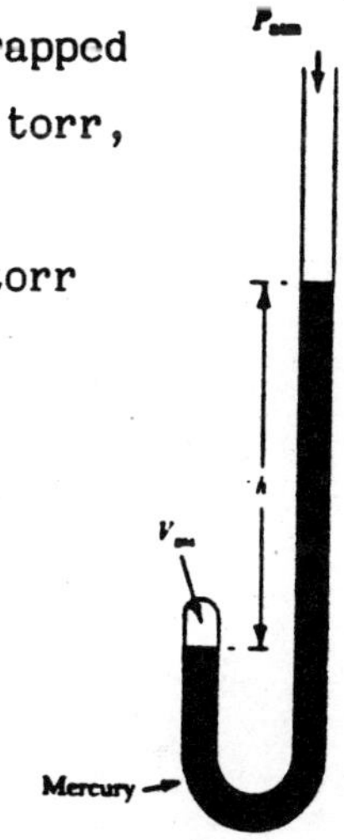

7.1 (c) 273-196 = 77 K

92% 7.2 (a) 235-273 = -38°C

7.3 (c) °C = °K - 273

52% 7.4 (e) P = 760 torr + 45 torr

80% 7.5 (b) $P_{atm} = P_T - h$ = 760 torr - 50 torr

Gas Laws

7.6 Which of the following scientists is credited with the statement, "at constant temperature, the volume occupied by a given mass of gas is inversely proportional to the pressure?

a. Avogadro b. Dalton c. Boyle d. Charles e. Graham

7.7 Who developed the following relationship, "at constant pressure, the volume of a gas is directly proportional to the absolute temperature?

a. Avogadro b. Dalton c. Boyle d. Charles e. Graham

7.8 Which of the graphs below represents the pressure-volume relationship between a given mass of gas at constant temperature?

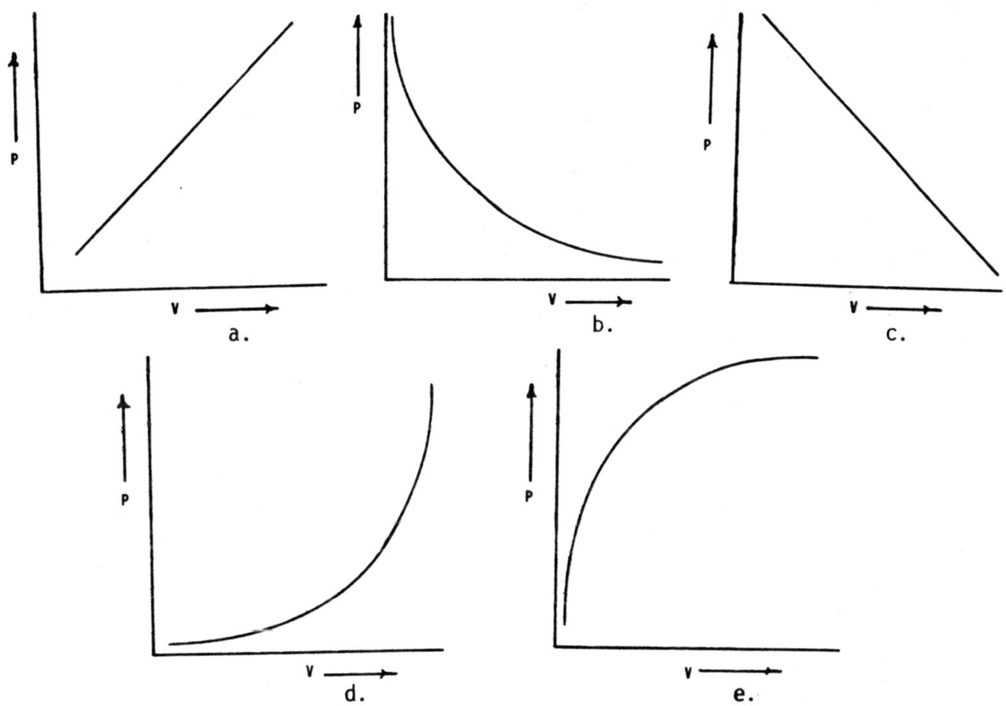

7.9 Absolute zero is the temperature at which

a. gaseous helium liquefies.

b. a straight line graph of P versus 1/V intersects the 1/V axis.

c. a straight line graph of V versus 1/P intersects the 1/P axis.

d. a graph of P versus V at constant T intersects the V axis.

e. a straight line graph of P versus T intersects the T axis.

7.6 (c) $V = k (1/P)$

7.7 (d) $V = kT$

7.8 (b) **At high pressures the volume is small and at low pressures the volume would be large.**

60% 7.9 (e) $PV = nRT$

$\therefore P \alpha T$ (a linear relationship)

P will be zero when T is zero.

7.10 For an ideal gas, which pair of variables are inversely proportional to each other (if all other factors remain constant)?

a. P, V   b. P, T   c. V, T   d. n, P   e. n, V

7.11 A weather balloon filled with hydrogen greatly increases in volume as it ascends to high altitudes. The expansion of the balloon is primarily due to

a. a decrease in the average kinetic energy of the hydrogen molecules inside the balloon.
b. an increase in the average kinetic energy of the hydrogen molecules inside the balloon.
c. a decrease in the average kinetic energy of the atmosphere molecules.
d. an increase in the rate of collisions of the hydrogen molecules against the inside walls of the balloon.
e. a decrease in the rate of collisions of the atmosphere molecules on the outside walls of the balloon.

7.12 A given mass of gas in a rigid container is heated from 300 K to 900 K. Which of the following responses best describes what will happen to the pressure of the gas?

a. The pressure will remain the same.
b. The pressure will decrease by a factor of three.
c. The pressure will increase by a factor of three.
d. The pressure will increase by a factor less than three.
e. The pressure will decrease by a factor greater than three.

7.13 A given mass of gas in a rigid container is heated from 100°C to 500°C. Which of the following responses best describes what will happen to the pressure of the gas?

a. The pressure will remain the same.
b. The pressure will decrease by a factor of five.
c. The pressure will increase by a factor of five.
d. The pressure will increase by a factor less than five.
e. The pressure will increase by a factor greater than five.

76% 7.10 (a) $PV = nRT$

$P = nRT\ (\frac{1}{V})$

$P \alpha \frac{1}{V}$

72% 7.11 (e) The kinetic energy of the hydrogen molecules and the molecules in the atmosphere must be the same since the molecules are all at the same temperature. There will be a decrease in the rate that hydrogen molecules are colliding with the inside walls of the balloon because of a decrease in temperature. The statement which is true is response (e).

68% 7.12 (e) $P_N = P_o$ x temperature correction factor.

$$P_N = P_o \times \frac{900}{300} = 3\ P_o$$

56% 7.13 (d) Remember to convert the temperatures to the Kelvin scale.

$P_N = P_o$ x temperature factor x volume factor.

$$P_N = P_o \times \frac{773}{373} \times \frac{1}{1} = 2.1\ P_o$$

7.14 A given mass of gas occupies the volume of 5.00 L at 65°C and 480 mm Hg. Which mathematical expression represents the volume of the gas at 630 mm Hg and 85°C?

a. $5.00 \times \frac{65}{85} \times \frac{480}{630}$ b. $5.00 \times \frac{338}{358} \times \frac{480}{630}$ c. $5.00 \times \frac{358}{338} \times \frac{480}{630}$

d. $5.00 \times \frac{358}{338} \times \frac{630}{480}$ e. $5.00 \times \frac{338}{358} \times \frac{630}{480}$

7.15 A gas occupies a volume of 1.50 L at 400 mm Hg and 100°C. Which mathematical expression gives the correct volume at 700 mm Hg and 200°C?

a. $1.50 \times \frac{400}{700} \times \frac{373}{473}$ b. $1.50 \times \frac{400}{700} \times \frac{473}{373}$ c. $1.50 \times \frac{700}{400} \times \frac{373}{473}$

d. $1.50 \times \frac{700}{400} \times \frac{473}{373}$ e. $1.50 \times \frac{400}{700} \times \frac{200}{100}$

7.16 A gas occupies a volume of 1.25 liters at 800 mm of Hg and 50.0°C. Which mathematical expression determines its volume at STP?

a. $1.25 \times \frac{800}{760} \times \frac{273}{323}$ b. $1.25 \times \frac{760}{800} \times \frac{273}{323}$ c. $1.25 \times \frac{760}{800} \times \frac{323}{273}$

d. $1.25 \times \frac{800}{760} \times \frac{50.0}{273}$ e. $1.25 \times \frac{760}{800} \times \frac{273}{50.0}$

7.17 The pressure of 4.0 L of an ideal gas in a flexible container is decreased to one-third of its original pressure and its absolute temperature is decreased by one-half. The volume then is

a. 1.0 L b. 4.0 L c. 6.0 L d. 8.0 L e. 24 L

80% 7.14 (c) $P_N = P_O$ x temperature factor x pressure factor

$$= 5.00 \times \frac{358}{338} \times \frac{480}{630}$$

The temperature factor must be greater than one. Why? The pressure factor must be less than one. Why?

92% 7.15 (b) $V_N = V_O$ x pressure factor x temperature factor

Why is the pressure factor less than one? Why is the temperature factor greater than one?

7.16 (a) At STP the pressure is 760 mm Hg which is less than 800 mm Hg. Therefore, multiply the volume by a pressure factor which is greater than one. Gas law calculations require that the temperature be in Kelvin degrees. At STP the temperature is lower than the original temperature so the temperature factor needs to be less than one.

7.17 (c) $V = 4.0\ L \times \frac{3}{1} \times \frac{1}{2} = 6.0\ L$

7.18 An ideal gas at 300 K and 760 mm Hg is contained in a flexible vessel and its volume is halved and the pressure remains unchanged. The temperature is

a. 75 K b. 150 K c. 300 K d. 600 K e. 1200 K

7.19 Complete combustion of a 0.20 mol sample of a hydrocarbon, C_xH_y, gives 0.80 mol of CO_2 and 1.0 mol of H_2O. The molecular formula of the original hydrocarbon is

a. C_3H_8 b. C_4H_5 c. C_4H_8 d. C_4H_{10} e. C_8H_{20}

7.20 When ammonium nitrite undergoes decomposition, only gases are produced according to the equation:

$$NH_4NO_2(s) \rightarrow N_2(g) + 2H_2O(g)$$

What is the total volume of gases produced at 819 K and 1.00 atm pressure when 128 g of ammonium nitrite undergoes the above decomposition reaction?

a. 3 x 22.4 L b. 6 x 22.4 L c. 9 x 22.4 L d. 12 x 22.4 L
e. 18 x 22.4 L

7.21 What is the molecular mass of a gas if 1.00 L of it at 0°C and 1.00 atm pressure has a mass of 2.86 g?

a. 22.4 g/mol b. 58.7 g/mol c. 64.1 g/mol d. 69.9 g/mol e. 80.2 g/mol

7.22 At standard conditions, it was found that 1.12 L of a gas weighed 2.78 g. Its molecular weight is

a. 2.78 g/mol b. 27.8 g/mol c. 55.6 g/mol d. 111 g/mol
e. Not enough information is given to solve this problem.

7.23 At 25°C and one atmosphere pressure, it is found that 2.24 L of gas weighs 3.11 g. Its relative molecular mass is

a. 3.11 g/mol b. 28.5 g/mol c. 31.1 g/mol d. 33.9 g/mol e. 37.1 g/mol

7.24 At 27°C and 1.00 atm, the density of a gaseous hydrocarbon is 1.79 $g \cdot L^{-1}$. The hydrocarbon is

a. CH_4 b. C_2H_4 c. C_2H_6 d. C_3H_8 e. C_4H_8

7.18 (b) $T_N = T_o$ x volume factor x pressure factor = $300 \times \frac{1}{2} \times \frac{1}{1}$ = 150 K
In order to have the same pressure, the molecules need to have less energy since there are now more of them per unit volume.

68% 7.19 (d) $\frac{\# C}{C_xH_y} = \frac{0.8 \text{ mol } CO_2}{0.2 \text{ mol } C_xH_y} \times \frac{1\ C}{1\ CO_2} = \frac{4\ C}{C_xH_y}$

$$\frac{\# H}{C_xH_y} = \frac{1.0 \text{ mol } H_2O}{0.2 \text{ mol } C_xH_y} \times \frac{2\ H}{1\ H_2O} = \frac{10\ H}{C_xH_y}$$

$\therefore C_4H_{10}$

7.20 (e) # mol NH_4NO_2 = 128 g $NH_4NO_2 \times \frac{1 \text{ mol}}{64 \text{ g } NH_4NO_2}$ = 2.0

$$= 2.0 \text{ mol } NH_4NO_2 \times \frac{3 \text{ mol gas}}{1\ NH_4NO_2} \times \frac{22.4 \text{ L}}{\text{mol}} \times \frac{819 \text{ K}}{273 \text{ K}} = 18 \times 22.4 \text{ L}$$

7.21 (c) # mol gas = $\frac{1 \text{ mol}}{22.4 \text{ L}} \times 1.00 \text{ L} = 0.0446$

$$\text{M.W.} = \frac{2.86 \text{ g}}{0.0446 \text{ mol}} = 64.1 \text{ g/mol}$$

7.22 (c) One mole = one gram molecular mass = 22.4 L at STP.

$$\frac{2.78 \text{ g}}{1.12 \text{ L}} \times \frac{22.4 \text{ L}}{1 \text{ mol}} = 55.6 \text{ g/mol}$$

56% 7.23 (d) $V_N = 2.24 \text{ L} \times \frac{273}{298} = 2.05 \text{ L}$

$$\frac{3.11 \text{ g}}{2.05 \text{ L}} \times \frac{22.4 \text{ L}}{1 \text{ mol}} = 33.9 \text{ g/mol}$$

or use the ideal gas equation.

$$MW = \frac{wt\ RT}{PV} = \frac{3.11 \text{ g} \times 0.08205 \frac{\text{L·atm}}{\text{mol·deg}} \times 298 \text{ deg}}{1.00 \text{ atm} \times 2.24 \text{ L}}$$

72% 7.24 (d) $MW = \frac{wt\ RT}{PV} = \frac{1.79 \text{ g} \times 0.08205 \text{ L·atm} \times 300 \text{ K}}{1.00 \text{ atm} \times 1.00 \text{ L}} = 44.1 \text{ g/mol}$

$MW_{C_3H_8} = 44.1$ g/mol

7.25 One mole of hydrogen, H_2, occupies 61.2 L at

a. 100°C and 1.00 atm. b. 200°C and 1.00 atm.

c. 0°C and 0.500 atm. d. 50°C and 0.500 atm.

e. 100°C and 0.500 atm.

7.26 A mixture of the gases, nitrogen, oxygen and carbon dioxide at 27°C and 0.50 atmosphere pressure occupied a volume of 492 ml. How many moles of gas are there in this sample?

a. 0.010 b. 1/9 c. 7.6 d. 10

e. Cannot be determined since it is a mixture.

7.27 A mixture consisting of 0.100 moles of N_2, 0.050 moles of O_2 and 0.200 moles of CH_4 and an unknown amount of CO_2 occupied a volume of 9.60 L at 27°Cdand 1.00 atmosphere pressure. How many moles of CO_2 are there in this sample?

a. 0.040 moles b. 0.050 moles c. 0.100 moles d. 0.390 moles

e. Cannot be determined since it is a mixture.

7.28 The density of N_2 gas at 0°C and 2.00 atmospheres is

a. 28 g/L b. 2.50 g/L c. 1.25 g/L d. 0.625 g/L

e. 0.028 g/ml

7.25 (e) The best approach for this problem is probably to try the different conditions that are listed and see which one gives 61.2 L.

$$1 \text{ mol } H_2 \times \frac{22.4 \text{ L}}{1 \text{ mol}} \times \frac{373}{273} \times \frac{1 \text{ atm}}{0.500 \text{ atm}} = 61.2 \text{ L}$$

7.26 (a) Use the equation PV = nRT. It doesn't matter what the gas molecules are.

$$n = \frac{PV}{RT} = \frac{0.50 \text{ atm} \times 0.492 \text{ L}}{0.08205 \frac{\text{L}\cdot\text{atm}}{\text{mol}\cdot\text{K}} \times 300 \text{ K}}$$

7.27 (a) $$n = \frac{PV}{RT} = \frac{1.00 \text{ atm} \times 9.60 \text{ L}}{0.08205 \frac{\text{L}\cdot\text{atm}}{\text{mol}\cdot\text{K}} \times 300 \text{ K}} = 0.390 \text{ mol}$$

$$\# \text{ mol } CO_2 = 0.390 - (0.100 + 0.050 + 0.200)$$
$$= 0.390 - 0.35 = 0.040$$

7.28 (b) $$d = \frac{P\cdot MW}{RT} = \frac{2.00 \text{ atm} \times 28.0 \text{ g/mol}}{0.08205 \frac{\text{L}\cdot\text{atm}}{\text{mol}\cdot\text{K}} \times 273 \text{ K}} = 2.50 \frac{\text{g}}{\text{L}}$$

$1 \text{ mol } N_2 = 22.4 \text{ L}$ at 0°C and 1.00 atm

$$V = 22.4 \text{ L} \times \frac{1.00 \text{ atm}}{2.00 \text{ atm}} = 11.2 \text{ L}$$

$$d = \frac{wt}{V} = \frac{28.0 \text{ g } N_2}{11.2 \text{ L}} = 2.50 \frac{\text{g}}{\text{L}}$$

Kinetic and Molecular Theory

7.29 Which of the following statements is LEAST likely to be true of a sample of nitrogen gas?

a. Molecules of gaseous nitrogen are in constant random motion.

b. The pressure exerted by gaseous nitrogen is due to collisions of the molecules with the walls of the container.

c. The average kinetic energy of the gaseous nitrogen is proportional to the absolute temperature of the gas.

d. Collisions between the gaseous molecules are elastic.

e. The volume of the sample would be zero at -273°C.

7.30 The molar volume of an ideal gas

a. is decreased by increasing the temperature.

b. always has a value of 22.4 L.

c. is decreased by decreasing the pressure of the gas.

d. is always lower than that of a real gas.

e. is decreased by decreasing the kinetic energy of the gas molecules.

7.31 In a sample of gas molecules at 298 K

a. all the molecules have the same velocity.

b. all the molecules have the same kinetic energy.

c. some of the molecules do not have any energy.

d. a few of the molecules have low velocities.

e. a few of the molecules have most of the kinetic energy.

68% 7.29 (e) Nitrogen has mass and even though it will be frozen at absolute zero, it must of necessity have a volume.

7.30 (e) It only has a value of 22.4 L at 273 K and 1 atm pressure. The volume decreases with a decrease in temperature.

7.31 (d) Also a few of the molecules have very high velocities.

Questions 7.32 and 7.33 refer to the information given below.

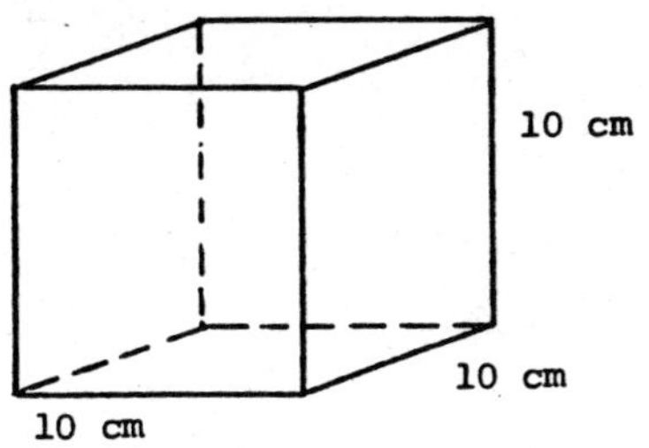

SO_2 at 25°C and
1.01 x 10^2 kPa pressure

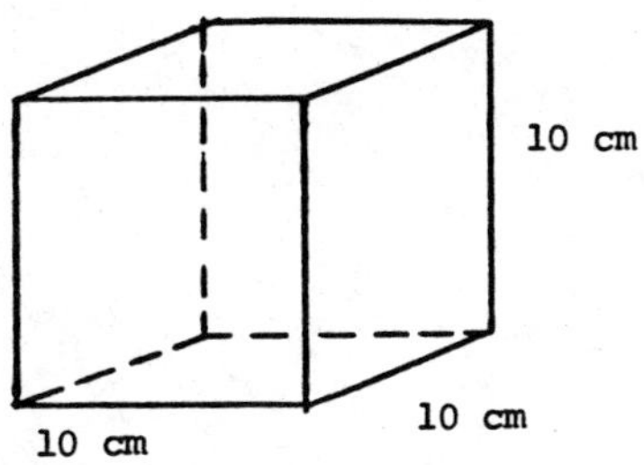

O_2 at 25°C and
1.01 x 10^2 kPa pressure

7.32 The ratio of the number of molecules of sulfur dioxide to the number of molecules of oxygen in each of the containers shown above is

a. 1 : 1 b. 1 : 2 c. 2 : 1 d. 3 : 2 e. $\sqrt{64 : 32}$

7.33 The <u>molecules</u> of the gases in the containers above have the same average

a. speed b. kinetic energy c. molecular mass

d. molecular volume e. van der Waals' gas constant a

7.34 According to the postulates of the kinetic theory of gases, the average speed of the molecules of a given gas is proportional to the

a. absolute temperature squared.

b. square root of the absolute temperature.

c. absolute temperature.

d. Celsius temperature squared.

e. reciprocal of the absolute temperature.

88% 7.32 (a) This question is testing a student's knowledge of Avogadro's law, at the same temperature and pressure equal volumes of all gases contain the same number of molecules.

80% 7.33 (b) The kinetic energy of a gas depends upon temperature. At the same temperature, all gases have the same average kinetic energy.

7.34 (b) K.E α $T^{1/2}$

7.35 At 320 K and 16 atm pressure, the molar of volume of ammonia, NH_3, is about 10% less than the molar volume of an ideal gas. The best explanation for the actual volume being this much smaller than the ideal volume is that

a. the volume occupied by the NH_3 molecules themselves is significant at this high of a concentration.

b. at this high temperature, a significant amount of NH_3 decomposes to N_2 and H_2.

c. the intermolecular forces of attraction become significant at this low of a temperature and this high of a pressure.

d. ammonia is a real gas and not an ideal gas.

e. the critical temperature and pressure of NH_3 (405 K and 112 atmospheres) are too close to the actual temperature and pressure of the NH_3 above.

7.36 A real gas most closely approaches ideal behavior at

a. 1.0 atm and 273 K. b. 10 atm and 546 K. c. 10 atm and 273 K.
d. 0.5 atm and 546 K. e. 0.5 atm and 273 K.

7.37 The behavior of HF(g) is most likely to approach the behavior of an ideal gas

a. when the temperature is 0°C and the pressure is one atmosphere.

b. when the temperature is decreased and the pressure is increased.

c. when the temperature is increased and the pressure is decreased.

d. when both temperature and pressure are increased.

e. when both temperature and pressure are decreased.

65% 7.35 (c) Through intermolecular hydrogen bonding, there are effectively fewer discrete molecules, when the molecules get close together.

60% 7.36 (d) Intermolecular forces become less significant when the kinetic energies of the molecules are relatively large. Intermolecular forces are not as effective at low concentrations (low pressures) as the molecules on the average are further apart.

72% 7.37 (c) Most real gases approach ideal behavior at high temperatures and at low concentrations.

7.38 If a sample of nitrogen gas in a sealed container of fixed volume is heated from 25°C to 250°C, the value of which of the following quantities will remain constant?

a. The average intensity of a molecular collision with the walls of the container.
b. The pressure of the gas.
c. The average speed of the molecules.
d. The average kinetic energy of the molecules.
e. The density of the nitrogen.

7.39 Under what set of conditions would $H_2S(g)$ be expected to have the best ideal behavior?

a. High temperature and low pressure.
b. Low temperature and high pressure.
c. High temperature and high pressure.
d. Low temperature and low pressure.
e. Standard temperature and pressure.

7.40 In the van der Waals equation of state for 1 mole of gas,

$$(P + \frac{a}{V^2})(V-b) = RT$$

the effect of intermolecular forces is accounted for by

a. $P + \frac{a}{V^2}$ b. P c. V-b d. b e. $(P + \frac{a}{V^2})(V-b)$

7.41 At a given temperature and pressure, which of the following gases would be expected to show the greatest deviation from the ideal gas law?

a. NH_3 b. N_2 c. NO d. CH_4 e. O_2

68% 7.38 (e) $d = \frac{m}{V}$ The volume is fixed and the mass of the gas is constant, therefore, the density will not change.

7.39 (a) Low concentrations and high kinetic energies (high temperatures).

52% 7.40 (a) The pressure term changes because of intermolecular attractive forces. The space taken up by the molecule is accounted for by the volume term, V-b.

60% 7.41 (a) NH_3 is a highly polar molecule. The dipole-dipole interaction between NH_3 molecules is large enough that it is called hydrogen bonding.

7.42 Under what set of conditions would $SO_2(g)$ be expected to <u>deviate</u> the most from ideal behavior?

a. High temperature and low pressure.

b. Low temperature and high pressure.

c. High temperature and high pressure

d. Low temperature and low pressure.

e. Standard temperature and pressure.

7.43 Which of the following substances is most likely to exist as a gas at 25°C and 1-atm paressure?

a. CaO b. C_8H_{18} c. $SbCl_3$ d. LiCl e. B_2H_6

7.44 The ratio of the average rate of effusion of $SO_2(g)$ to $CH_4(g)$ at 300 K is

a. 1:4 b. 1:2 c. 4:1 d. 2:1 e. $\sqrt{64}$: $\overline{16}$

7.45 The ratio of the <u>average speeds</u> of $SO_2(g)$ to HBr(g) is

a. 8 to 9 b. 1 to 1 c. 9 to 8 d. 64 to 81 e. 81 to 64

7.46 The ratio of the average <u>kinetic energies</u> of $SO_2(g)$ to HBr(g) at 25°C is

a. 8 to 9 b. 1 to 1 c. 9 to 8 d. 64 to 81 e. 81 to 64

7.47 The ratio of the average kinetic energies of O_2 to CH_4 at 298 K is

a. 2 to 1 b. 1 to 1 c. 1 to 2 d. $\sqrt{32}$ to $\sqrt{16}$ e. $\sqrt{16}$ to $\sqrt{32}$

7.48 The ratio of the <u>average speeds</u> of O_2 to CH_4 at 298 K is

a. 2 to 1 b. 1 to 1 c. 1 to 2 d. $\sqrt{32}$ to $\sqrt{16}$ e. $\sqrt{16}$ to $\sqrt{32}$

7.49 If 250 ml of methane, CH_4, effuses through a small hole in 48 seconds, the time required for the same volume of helium to pass through the hole will be

a. 12 sec. b. 24 sec. c. 48 sec. d. 96 sec. e. 192 sec.

7.42 (b) When the average kinetic energy of the molecules is low (low temperature), the intermolecular forces become important. High concentrations (high pressures) contribute to the nonideality of the gases.

40% 7.43 (e) First, let us eliminate ionic type compounds (metals + nonmetals). Select a small covalent molecule.

7.44 (b) $\dfrac{\text{Rate } SO_2}{\text{Rate } CH_4} = \sqrt{\dfrac{\text{M.W. } CH_4}{\text{M.W. } SO_2}} = \sqrt{\dfrac{16}{64}}$

64% 7.45 (c) $\dfrac{\text{Rate } SO_2}{\text{Rate HBr}} = \sqrt{\dfrac{81}{64}}$

80% 7.46 (b) K.E. = $1/2\ mv^2$. The kinetic energy changes with (and is proportional to) the absolute temperature.

7.47 (b) See the answer to 7.46 above.

7.48 (e) Remember the inverse relationship.

7.49 (b) This effusion experiment involves <u>time</u> not rate.

$$\frac{\text{Rate He}}{\text{Rate } CH_4} = \sqrt{\frac{\text{M.W. } CH_4}{\text{M.W. He}}} = \frac{t_{CH_4}}{t_{He}}$$

$$\sqrt{\frac{16}{4}} = \frac{48}{t_{He}} \quad ; \quad t_{He} = 24 \text{ s}$$

7.50 The molecular weight of an unknown gas was measured by an effusion experiment. It was found that it took 82 secs. for the gas to effuse whereas oxygen required 55 secs. The molecular weight of the gas is

a. 21 g/mol b. 35 g/mol c. 71 g/mol d. 84 g/mol

e. Cannot be determined from these data.

7.51 It takes 125 seconds for a 10.0 mL sample of an unknown gas to effuse through a pinhole. A 10.0 mL sample of oxygen takes 94 seconds. What is the molecular mass of the unknown gas?

a. 18.1 g/mol b. 28.3 g/mol c. 48.2 g/mol d. 56.6 g/mol

e. 74.3 g/mol

7.50 (c) Remember time has been given in this experiment not rate.

$$\frac{t_u}{t_{O_2}} = \sqrt{\frac{M.W._u}{M.W._{O_2}}} \quad ; \; M.W._u = \frac{82^2 \times 32}{55^2}$$

7.51 (d) Time is directly proportional to the molecular weight.

$$M.W._u = \frac{125^2 \times 32}{94^2} = 56.6 \text{ g/mol}$$

8

LIQUIDS AND SOLIDS

Solid Structure

8.1 Which of the following as solids has a crystal structure containing discrete (or separate) molecules?

a. potassium b. glass c. quartz d. carborundum, SiC

e. hydrogen

8.2 An atom that is shared equally between eight cubic unit cells is called

a. a corner atom. b. a face atom. c. a body-centered atom.

d. an edge atom. e. a diagonal atom.

8.3 Europium crystallizes in a body-centered cubic lattice. The number of europium atoms in a unit cell is

a. 1 b. 2 c. 4 d. 6 e. 8

8.4 A crystalline compound consists of two types of atoms, X and Y. It crystallizes in a cubic lattice with one X atom at the body center and Y atoms at the corners of the unit cell. The simplest formula of this compound is

a. XY_8 b. X_2Y c. XY d. XY_2 e. X_8Y

8.5 Tungsten crystallizes in a cubic lattice with two atoms of W per unit cell. It could be classified as

a. a body-centered cubic structure.

b. a face-centered cubic structure.

c. a simple cubic structure.

d. a cubic closest packing structure.

e. a hexagonal closest packing structure.

8.1 (e) Potassium has a metallic bonding (no discrete metal atoms). Glass and quartz both consist of giant SiO_2 molecules. Silicon carbide has a 3-dimensional diamond like structure and it is a giant molecule. The intermolecular forces between the hydrogen molecules are van der Waal type interactions.

8.2 (a) An edge atom is shared between four unit cells. A face atom is shared between two unit cells. A body-centered atom is shared by one unit cell.

8.3 (b) # atoms = 8 x 1/8 + 1 X 1 = 2

92% 8.4 (c) # X atoms = 1 X 1 = 1

Y atoms = 1 X 1/8 = 1

76% 8.5 (a) 8 x 1/8 + 1 X 1 = 2

8.6 What is the simplest formula of a solid containing A, B, and C atoms in a cubic lattice in which the A atoms occupy the corners, the B atoms, the body-centered position and the C atoms, the faces of the unit cell?

a. ABC b. ABC_3 c. ABC_6 d. A_8BC_6 e. A_4BC_3

8.7 An ionic crystalline solid, MX_2, has a cubic unit cell. Which of the following arrangements of the ions is consistent with the stoichiometry of the compound?

a. M^{2+} ions at the corners and at each face, 8 X^- ions at the body centers.

b. M^{2+} ions at each face, X^- ions at the corners.

c. M^{2+} ions at the corners, X^- ions at the faces and 2 X^- ions at the body centers.

d. 4 M^{2+} ions at the body centers, X^- ions at the corners and at the face centers.

e. M^{2+} ions at the corners and at the face centers, 4 X^- ions at the body centers.

8.8 Solid nickel has a cubic unit cell. The edge of the cubic cell is 3.0×10^{-8} cm. Which of the cubic unit cells described below would give nickel its highest density?

a. A simple cubic cell.

b. A body-centered cubic cell.

c. A face-centered cubic cell.

d. The density would be identical for all of the cells above if the edge of each cell was 3.0×10^{-8} cm.

e. Cannot determine since the size of the nickel atom is not given.

8.9 The oxide ions in Na_2O form a face-centered cubic lattice and the sodium ions occupy all the tetrahedral holes. The oxide ions occupy the faces and the corners of the unit cell. How many sodium ions are in one unit cell?

a. 2 b. 4 c. 6 d. 8 e. 12

8.6 (b) # A = 8 x 1/8 = 1

B = 1 x 1 = 1

C = 6 x 1/2 = 3

60% 8.7 (a) # M^{2+} = 8 x 1/8 + 6 x 1/2 = 4

X^- = 8 x 1 = 8

∴ MX_2

8.8 (c) A face-centered cubic has more atoms per unit cell. A simple cubic structure has one atom per unit cell. A body-centered cubic structure has two atoms per unit cell. A face-centered cubic structure has four atoms per unit cell.

8.9 (d) There are 2 Na^+/1 O^{2-}

O^{2-} = 8 x 1/8 + 6 x 1/2 = 4

8.10 In UO_2, the uranium ions form a face-centered cubic lattice. The oxide ions occupy the tetrahedral sites. The uranium ions occupy the faces and corners of the unit cell. The <u>total number</u> of uranium ions and oxide ions per unit cell is

a. 2 b. 4 c. 6 d. 8 e. 12

8.11 Titanium dioxide crystallizes in a non cubic crystal system (a tetragonal system). In this structure, two of the O^{2-} ions are within the interior of the cell, two are in the top face, and two in the bottom face of the cell. Ti^{4+} ions are at the corners and the center of the cell. How many oxide ions are there per unit cell?

a. 1 b. 2 c. 3 d. 4 e. 6

8.12 Silver crystallizes in the face-centered cubic system. If the edge of the unit cell is 4.07 Å, what is the radius of a silver atom in Å (1 Å $= 10^{-8}$ cm)?

a. 1.44 b. 1.76 c. 2.04 d. 2.88 e. 3.52

8.13 Which of the following as <u>solids</u> has a crystal structure containing <u>discrete</u> (or separate) molecules?

a. NaCl b. CH_4 c. graphite d. diamond e. SiO_2

8.14 A solid has a very high melting point, it is very hard, and its liquid is nonconducting. The solid is

a. an ionic solid b. a covalent solid c. a metallic solid
d. an atomic solid e. a molecular solid

8.15 Which of the following compounds would be expected to have the highest melting point?

a. LiCl b. CCl_4 c. $MgCl_2$ d. NCl_3 e. OCl_2

8.10 (e) # U^{4+} ions = 6 x 1/2 + 8 x 1/8 = 4

$$\# \, O^{2-} \text{ ions} = \frac{2\,O^{2-}}{1\,U^{4+}} \times 4\,U^{4+} = 8$$

76% 8.11 (d) # O^{2-} = 2 X 1 + 2 X 1/2 + 2 X 1/2 = 4

Ti^{4+} = 8 X 1/8 + 1 X 1 = 2

$$\# \, O^{2-} = \frac{2\,O^{2-}}{1\,Ti^{4+}} \times 2\,Ti^{4+} = 4$$

8.12 (a)

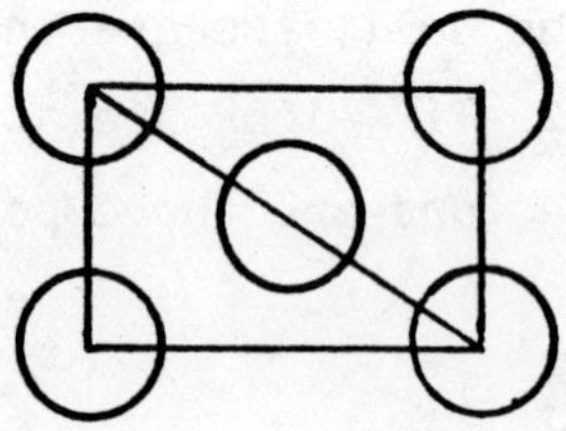

$e^2 + e^2 = FD^2$

$FD^2 = 2 \times (4.07)^2$

FD = 5.75 = 4 radii

$$1 \text{ radius} = \frac{5.75\ \text{Å}}{4} = 1.44\ \text{Å}$$

8.13 (b) Graphite, diamond and SiO_2 are giant molecules. Sodium chloride has a crystalline lattice that consists of an infinite array of ions. Methane molecules in the solid state are held together by weak van der Waal type forces.

72% 8.14 (b) The solid is probably a large macromolecule such as silicon carbide, diamond or silicon dioxide.

56% 8.15 (c) In this question we only need to consider LiCl and $MgCl_2$ because CCl_4, NCl_3 and OCl_2 are covalent compounds. The ionic compound (LiCl or $MgCl_2$) with the higher melting point will be the one with the larger crystal lattice energy.

$$U = \frac{Mz^+z^-e^2}{r_0}$$

where z^+ and Z^- are the charges on the positive ions and negative ions respectively and r_o is the distance that they are separated. The smaller the ions the smaller r_o will be and the higher U will be. M is a geometric factor that is characteristic for different types of crystals.

8.16 Which of the following compounds would be expected to have the highest melting point?

a. LiCl b. NaCl c. $MgCl_2$ d. $BaCl_2$ e. BCl_3

8.17 Which of the following compounds would be expected to have the highest melting point?

a. NaF b. AlF_3 c. SiF_4 d. PF_5 e. SF_6

Phase Diagrams and Phase Changes

8.18 In the phase diagram below, the solid, liquid and vapor phases are represented respectively by

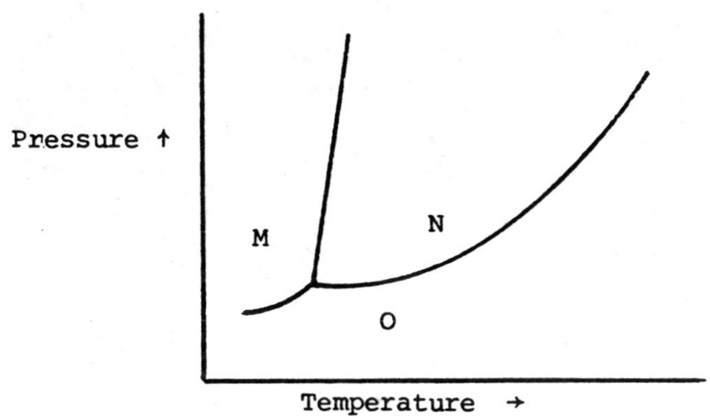

a. O, M, N b. O, N, M c. N, O, M d. N, M, O e. M, N, O

30% 8.16 (c) We should only consider the ionic compounds LiCl, NaCl and $MgCl_2$. $MgCl_2$ should have the highest crystal lattice energy because of the charge on the Mg^{2+} ion. See the answer to question 8.15 above.

30% 8.17 (b) AlF_3 will have the largest crystal lattice energy because of the high charge on the aluminum ion. In this question we only need to consider NaF and AlF_3 as the other compounds are covalent so the intermolecular forces are van der waal type interactions.

8.18 (e) O will be vapor because it exists under low pressures and high temperatures. M will be the solid state as substances will exist as solids at low temperatures and high pressures.

8.19 From a consideration of the phase diagram below, a change from point M to point N corresponds to

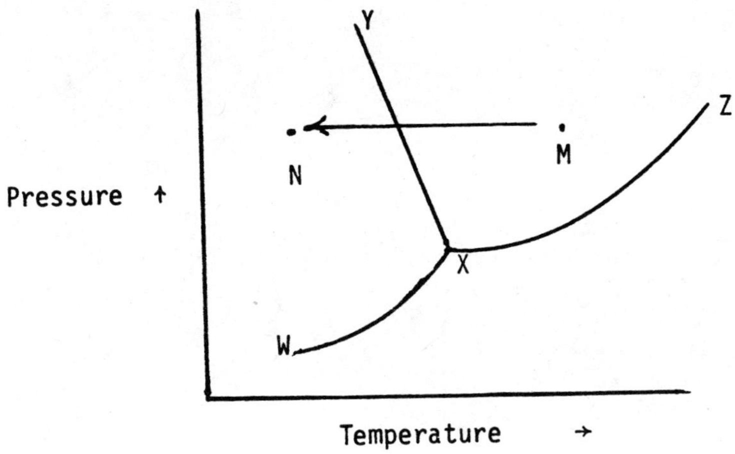

a. sublimation b. condensation c. evaporation d. solidification
e. liquefaction

8.20 The lines in the diagram represent values of pressure and temperature at which two phases may be present in equilibrium. Which one of the following descriptions of the lines SX, RX, and XZ is the correct one?

a. SX = solid-liquid, RX = solid-gas, XZ = gas-liquid.
b. SX = solid-gas, RX = solid-liquid, XZ = gas-liquid.
c. SX = solid-liquid, RX = gas-liquid, XZ = solid-gas.
d. SX = gas-liquid, RX = solid-gas, XZ = solid-liquid.
e. SX = gas-liquid, RX = solid-liquid, XZ = solid-gas.

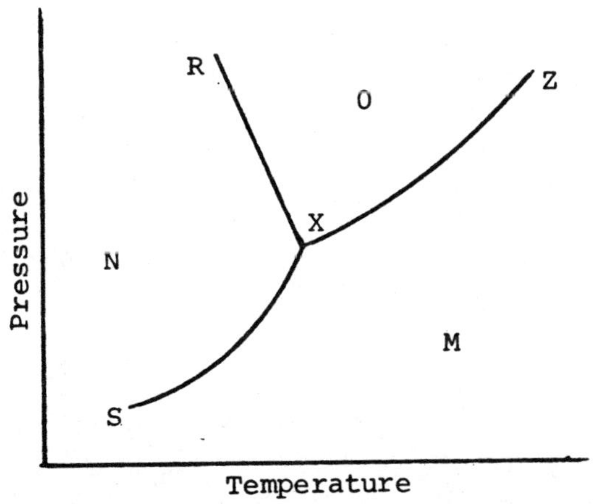

92% 8.19 (d) The change of state is from liquid to solid.

8.20 (b) N represents the solid area, O represents the liquid area and M the gaseous area.

8.21 From a consideration of the phase diagram below, a change from point N to point M corresponds to

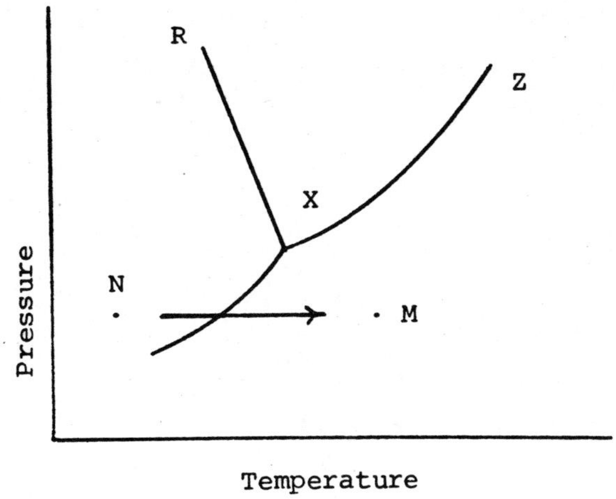

a. sublimation b. vaporization c. condensation d. liquefaction
e. freezing

8.22 The sublimation of solid carbon dioxide, dry ice, is an example of
a. evaporation b. a physical property c. a chemical property
d. a chemical change e. a physical change

8.23 The process which is represented by the equation, K(s) → K(g) is
a. fusion b. melting c. sublimation d. vaporization
e. condensation

8.24 The heat of sublimation of a compound equals
a. heat of fusion plus heat of vaporization.
b. heat of ionization plus heat of crystallization.
c. heat of vaporization minus heat of fusion.
d. heat of vaporization plus heat of crystallization.
e. heat of crystallization minus heat of vaporization.

92% 8.21 (a) Point N is in the solid state and Point M is in the gaseous state.

8.22 (e) Change in states correspond to physical changes.

8.23 (c) A substance going directly from the solid state to the gaseous undergoes sublimation.

8.24 (a)

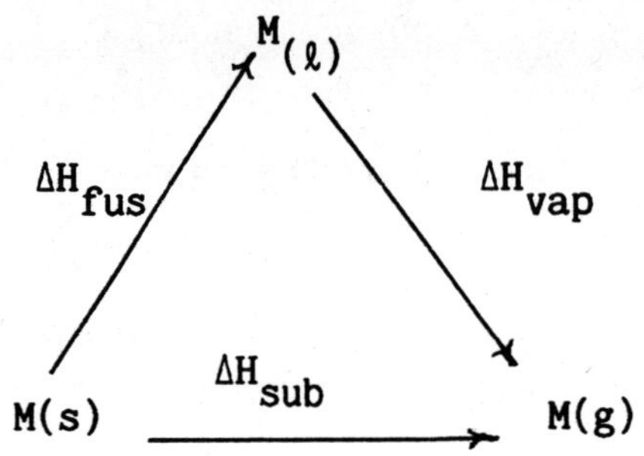

$$\Delta H_{sub} = \Delta H_{fus} + \Delta H_{vap}$$

8.25 The heat of fusion of a compound equals

a. heat of vaporization plus heat of crystallization.

b. heat of fusion plus heat of vaporization.

c. heat of sublimation minus heat of vaporization.

d. heat of vaporization minus heat of fusion.

e. heat of ionization plus heat of crystallization.

8.26 The critical point of carbon tetrachloride is 283°C and 45 atm pressure. Liquid carbon tetrachloride has a vapor pressure of 10.0 atm at 178°C. Which of the following statements must be true?

a. The normal boiling of CCl_4 must be greater than 178°C.

b. Liquid CCl_4 can exist at temperatures greater than 283°C if the pressure is greater than 45 atm.

c. The triple point must be less than 178°C.

d. Liquid and solid can only be in equilibrium at one temperature - the freezing point.

e. Vapor and liquid can only be in equilibrium at one temperature - the normal boiling point.

Vapor Pressure and Boiling Point

8.27 If heat is added to ice and liquid water in a closed container, and after the addition of the heat there is still ice and liquid water remaining

a. the vapor pressure of water will decrease.

b. the temperature will increase some.

c. the temperature will decrease some.

d. the vapor pressure of water will rise.

e. the vapor pressure of water will remain constant.

8.25 (c)

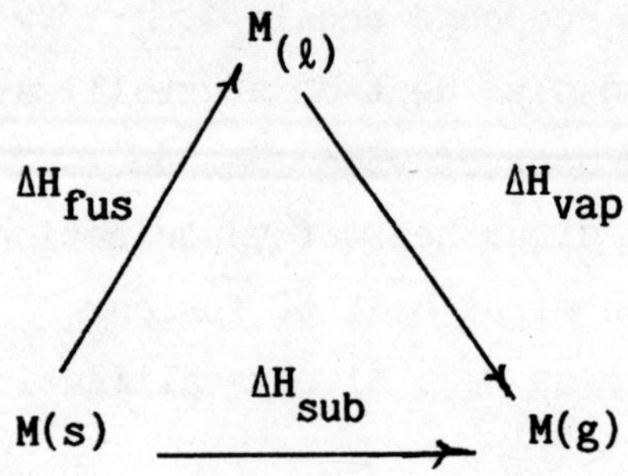

$$\Delta H_{fus} = \Delta H_{sub} - \Delta H_{vap}$$

8.26 (c) The triple point is when the liquid, solid and gaseous phases are in equilibrium. To have solid CCl_4, the temperature must be less than that at the critical point because at that point, there are present just the liquid and gas phases.

8.27 (e) The temperature of the liquid will still be the same. Therefore, the vapor pressure of water will be the same.

8.28 A closed vessel contains an inert gas A at 500 mm of Hg and 60°C. What is the total pressure in mm of Hg inside the vessel after a few drops of liquid B are injected and a phase equilibrium is established at 60°C? The vapor pressure of liquid B at 60°C is 80 mm of Hg.

a. 420 b. 440 c. 500 d. 560 e. 580

8.29 The vapor pressure of a liquid increases with an increase of temperature. Which of the following statements best explains this increase?

a. The average kinetic energy of molecules is greater, thus more molecules can enter the gaseous state.
b. The number of gaseous molecules above the liquid remains constant but these molecules have greater average kinetic energy.
c. The faster moving molecules in the liquid exert a greater pressure.
d. All the molecules have greater kinetic energies.
e. The intermolecular forces between the molecules becomes less at higher temperatures.

8.30 The normal boiling point of a liquid

a. is above the critical temperature.
b. is at the triple point temperature.
c. is at one atmosphere pressure.
d. is usually below the triple point temperature.
e. is only when the liquid and vapor are in equilibrium.

8.31 The normal boiling point of a liquid

a. is 100°C at 1 atm pressure.
b. is the temperature at which the vapor pressure is 1 atm.
c. is the temperature at which liquid and vapor are in equilibrium.
d. is the temperature at which the vapor pressure equals the external pressure.
e. is the temperature at which there is a continuous formation of gaseous bubbles in the liquid.

8.28 (e) This represents an application of Dalton's law of partial pressures, $P_T = P_A + P_B \ldots$

56% 8.29 (a) More molecules have sufficient energy to escape from the liquid. The intermolecular forces between the molecules are just as strong at a higher temperature, but they become less important as the kinetic energy increases.

8.30 (c) It is important to know what is meant by boiling point and normal boiling point. See question 8.31 below.

72% 8.31 (b) Response (a) is only correct for water but all liquids have their own characteristic boiling points. Responses (d) and (e) define boiling but not the normal boiling point.

8.32 The vapor pressure of a given liquid will increase if

a. the liquid is moved to a container in which its surface is very much larger.

b. the volume of the liquid is increased.

c. the volume of the vapor phase is increased.

d. the temperature is increased.

e. a more volatile liquid is added to the given liquid.

8.33 When a small amount of heat is added to a pure liquid (a single compound) that is just starting to boil, what is the result?

a. The temperature decreases slightly.

b. The temperature remains the same.

c. The temperature increases slightly.

d. The liquid becomes superheated.

e. The liquid becomes supercooled.

8.34 The data for the equilibrium pressure of any liquid as a linear function of temperature can best be represented by plotting

a. log P versus t   b. log P versus T   c. log P versus 1/T

d. P versus t   e. P versus T

Intermolecular Forces in Liquids

8.35 Which of the following indicates very strong intermolecular forces of attraction in a liquid?

a. A very low boiling point.

b. A very low critical temperature.

c. A very low heat of vaporization.

d. A very low vapor pressure.

e. A very low surface tension.

88% 8.32 (d) As the kinetic energy of the molecules increases, the vapor pressure must increase. The kinetic energy of the molecules is proportional to the absolute temperature.

76% 8.33 (b) The added heat provides additional heat for more molecules to escape from the liquid state (ΔH_{vap}). The temperature will not change until all the liquid has changed to the gaseous state.

88% 8.34 (c) $\log P = \frac{-\Delta H_{vap}}{2.303\ R} \left(\frac{1}{T}\right) + C$

$y = \quad m\ x + b$

The equation above which illustrates vapor pressure dependence as a function of temperature is known as the Clausius-Clapeyron equation. From such a relationship (from the slope of a straight line) it is possible to obtain ΔH_{vap}, the heat of vaporization.

8.35 (d) A high vapor pressure is an indication of weak intermolecular forces.

8.36 All of the following indicate very strong intermolecular forces of attraction in a liquid EXCEPT

a. a very high boiling point.

b. a very high vapor pressure.

c. a very high critical temperature.

d. a very high viscosity.

e. a very high heat of vaporization.

8.37 The compounds Br_2 and ICl have almost identical molecular weights, yet ICl boils at 97°C and Br_2 boils at 59°C. The best explanation for the difference is

a. ICl is an ionic compound and Br_2 is covalent.

b. ICl is a nonpolar molecule and Br_2 is polar.

c. ICl has a longer bond than that in Br_2.

d. ICl has a measureable dipole moment and Br_2 does not.

e. ICl has a stronger bond than that in Br_2.

8.38 Which compound should have the highest vapor pressure at room temperature?

a. C_4H_{10} b. C_5H_{12} c. C_6H_{14} d. C_7H_{16} e. C_8H_{18}

8.39 In some compounds the hydrogen atom is covalently bonded to one atom and simultaneously attracted to another atom in another molecule by an electrostatic interaction. This interaction can occur when hydrogen is bonded to

a. Cl b. Si c. N d. C e. Br

8.40 Which of the following compounds would be expected to have the lowest boiling point?

a. $CH_3CH_2CH_2F$ b. $CH_3CH_2CH_2OH$ c. CH_3CHCH_3 (with OH on the central C) d. CH_3CH_2COOH

e. $CH_3CH_2CH_2NH_2$

56% 8.36 (b) A low vapor pressure is an indication of strong intermolecular forces of attraction between molecules.

8.37 (d) ICl is a polar molecule whereas Br_2 is a nonpolar molecule. There are dipole-dipole interactions between ICl molecules but only London dispersion forces between Br_2 molecules.

8.38 (a) Another question to ask is which compound has the lowest boiling point. The smallest molecule will have the lowest boling point, the weakest intermolecular forces.

8.39 (c) Hydrogen bonding can occur when hydrogen is bonded to the very electronegative atoms, N, O and F.

30% 8.40 (a) Many students think that $CH_3CH_2CH_2F$ has a high boiling point because of the fluorine atom. There is no hydrogen bonding in this molecule because the fluorine atom is not bonded to a hydrogen atom.

8.41 Which of the following compounds shows an abnormal boiling point due to hydrogen bonding?

a. CH_3NH_2 b. CH_3OCH_3 c. CH_3SH d. CH_3Cl e. HCl

8.42 Which of the following compounds is expected to have the highest boiling point?

a. CH_3CH_3 b. $CH_3CH_2CH_3$ c. CH_3OCH_3 d. $CH_3CH_2CH_2Cl$

e. $CH_3CH_2CH_2Br$

8.43 Of the following substances, the one expected to have the highest critical temperature is

a. CH_3OH b. CH_3Cl c. CH_4 d. C_2H_6 e. C_3H_8

8.44 Which of the following compounds is expected to have the highest boiling point?

a. $CH_3CH_2CH_3$ b. $CH_3CH_2CH_2CH_3$ c. CH_3OCH_3 d. CH_3CH_2OH

e. CH_3CH_2F

8.45 Which of the following would be expected to have the highest heat of vaporization?

a. H_2O b. NH_3 c. PH_3 d. AsH_3 e. SbH_3

8.46 Which of the following has the lowest boiling point?

a. H_2O b. H_2S c. H_2Se d. H_2Te e. NH_3

8.47 Which of the following compounds would be expected to have the highest boiling point?

a. $CH_3CH_2NH_2$ b. CH_3CH_2F c. CH_3OCH_3 d. CH_3CHO

e. $CH_3CH_2CH_3$

8.41 (a) In methylamine, CH_3NH_2, there is a hydrogen atom bonded to the nitrogen. In dimethylether, CH_3OCH_3, there is no hydrogen bonding because no hydrogen atoms are bonded to oxygen.

44% 8.42 (e) Which molecule is the largest or has the most electrons?

8.43 (a) Which compound has the highest boiling point? In which compound is there hydrogen bonding between the molecules?

30% 8.44 (d) There is hydrogen bonding between the ethanol molecules. When the molecules are very similar in size, then the molecules which have hydrogen bonding will have the highest boiling point. There is no hydrogen bonding with CH_3CH_2F.

8.45 (a) In which compound are the intermolecular forces the greatest? Which compound has the highest boiling point? Hydrogen bonding is more extensive in water than in ammonia.

8.46 (b) Which molecule is the smallest but does not have any hydrogen bonding? Both H_2O and NH_3 involve hydrogen bonding.

30% 8.47 (a) In which compound is hydrogen bonding operative as an intermolecular force? H-bonding between molecules of $CH_3CH_2NH_2$. No H-bonding in CH_3CH_2F because H is not bonded to F. No H-bonding in CH_3CHO because H is not bonded to O.

9

PROPERTIES OF SOLUTIONS

Molality and Mole Fraction

9.1 What is the mole fraction of water in a water-ethanol solution that is 54.0% water by mass? Ethanol is C_2H_5OH.

a. 0.25 b. 0.33 c. 0.46 d. 0.67 e. 0.75

9.2 What is the mole fraction of methanol, CH_3OH, in an ethanol, C_2H_5OH, solution that is 60.0% ethanol by mass?

a. 0.40 b. 0.46 c. 0.49 d. 0.54 e. 0.60

9.3 An aqueous solution of ethanol, C_2H_5OH, that is 12.0% ethanol by weight, has a density of 0.981 $g \cdot mL^{-1}$ at 20°C. The molecular mass of C_2H_5OH is 46.0 $g \cdot mol^{-1}$. What is the molality of ethanol in this solution?

a. 1.20 b. 1.33 c. 2.56 d. 2.66 e. 2.96

9.4 The molality of a solution is

a. moles of solute per liter of solution.
b. moles of solute per mole of solvent.
c. moles of solute per liter of solvent.
d. moles of solute per kilogram of solvent.
e. moles of solute per kilogram of solution.

9.5 A solution is prepared by dissolving 0.500 mole of Glauber's salt (sodium sulfate decahydrate), $Na_2SO_4 \cdot 10H_2O$ in 910 g of water. The water of hydration becomes part of the solvent. In terms of the Na_2SO_4, the solution has the concentration

a. 0.50 M b. 0.50 m c. 0.55 M d. 0.55 m e. 0.60 M

88% 9.1 (e) Assume 100. g of solution.

$$\# \text{ mol } H_2O = 54.0 \text{ g } H_2O \times \frac{1 \text{ mol } H_2O}{18.0 \text{ g } H_2O} = 3.00$$

$$\# \text{ mol } C_2H_5OH = 46.0 \text{ g } C_2H_5OH \times \frac{1 \text{ mol } C_2H_5OH}{46.0 \text{ g } C_2H_5OH} = 1.00$$

$$X_{H_2O} = \frac{n_{H_2O}}{n_{H_2O} + n_{C_2H_5OH}} = \frac{3}{3+1} = 0.75$$

9.2 (c)

$$X_{CH_3OH} = \frac{n_{CH_3OH}}{n_{CH_3OH} + n_{C_2H_5OH}}$$

$$= \frac{40.0 \text{ g} \times \frac{1 \text{ mol}}{32.0 \text{ g}}}{40.0 \text{ g} \times \frac{1 \text{ mol}}{32.0 \text{ g}} + 60.0 \text{ g} \times \frac{1 \text{ mol}}{46.0 \text{ g}}}$$

70% 9.3 (e) Assume 100.0 mL of solution.

$$wt_{solution} = 100.0 \text{ mL} \times 0.981 \text{ g/mL} = 98.1 \text{ g}$$

$$wt_{C_2H_5OH} = 98.1 \text{ g} \times 0.120 = 11.8 \text{ g}$$

$$\text{wt of } H_2O = 98.1 \text{ g} - 11.8 \text{ g} = 86.3 \text{ g}$$

$$m = \frac{n}{1.00 \text{ kg solvent}} = \frac{11.8 \text{ g } C_2H_5OH \times \frac{1 \text{ mol}}{46.0 \text{ g}}}{0.0863 \text{ kg}}$$

88% 9.4 (d)

$$m = \frac{n_{solute}}{1.00 \text{ kg solvent}}$$

30% 9.5 (b)

$$\# \text{ mol } Na_2SO_4 = \# \text{ mol } Na_2SO_4 \cdot 10\ H_2O$$

$$\# \text{ mol } H_2O \text{ from salt} = \frac{10\ H_2O}{1\ Na_2SO_4 \cdot 10\ H_2O} \times 0.500 \text{ mol}$$

$$\text{Total wt of } H_2O = 5.0 \text{ mol} \times \frac{18.0 \text{ g } H_2O}{1 \text{ mol } H_2O} + 910 \text{ g} = 1000 \text{ g}$$

$$m = \frac{0.500}{1.000} \frac{\text{mol}}{\text{kg}}$$

9.6 If 0.100 mole of napthalene is dissolved in 100. g of benzene (C_6H_6), the molality is

a. 0.100 b. 0.900 c. 1.00 d. 1.28 e. 12.8

9.7 A solution is prepared by dissolving 0.500 mole of $Al_2(SO_4)_3 \cdot 18H_2O$ in 838 g of water. The water of hydration becomes part of the solvent. In terms of the anhydrous salt ($Al_2(SO_4)_3$), the solution has the concentration

a. 0.500 m b. 0.600 m c. 0.500 M d. 0.600 M e. 0.400 m

Solutions

9.8 All of the following are colligative properties of a solution EXCEPT

a. freezing point depression.
b. boiling point elevation.
c. vapor pressure lowering.
d. osmotic pressure.
e. solute solubility.

9.9 Which property of a solution is <u>not</u> a colligative property?

a. Freezing-point lowering
b. Boiling-point elevation
c. Surface tension
d. Osmotic pressure
e. Vapor-pressure lowering

9.10 All of the following affect the solubility of gases in solvents EXCEPT

a. nature of the gas.
b. nature of the solvent.
c. pressure of the gas.
d. temperature.
e. atmospheric pressure.

9.6 (c) $m = \frac{\text{\# mol solute}}{1.00 \text{ kg solvent}} = \frac{0.100}{0.100} = 1.00$

25% 9.7 (a) Most of the students forgot the water of hydration which becomes part of the solvent.

\# mol H_2O in 0.500 mol $Al_2(SO_4)_3 \cdot 18\ H_2O$

$$0.500 \text{ mol } Al_2(SO_4)_3 \cdot 18\ H_2O \times \frac{18\ H_2O}{1\ Al_2(SO_4)_3 \cdot 18\ H_2O} = 9.00 \text{ mol}$$

$$\text{wt of } H_2O = 9.00 \text{ mol } H_2O \times \frac{18.0 \text{ g } H_2O}{1 \text{ mol } H_2O} + 838 \text{ g} = 1000 \text{ g}$$

$$m = \frac{0.500 \text{ mol}}{1.00 \text{ Kg}}$$

80% 9.8 (e) The solubility of a solute depends upon the nature of the solute. Every solute has its own characteristic solubility. A colligative property is independent of the kind of solute.

9.9 (c) A colligative property depends upon the number of solute particles and not the kind of particles. The four common colligative properties are given in questions 9.8 and 9.9.

9.10 (e) Henry's law is C(g) = kP where k is the constant at a given temperature for the gas and P is the partial pressure of the gas. The atmospheric pressure does not enter into this relationship. The constant k depends upon both nature of the gas and of the solvent.

9.11 The solubility of a gas in a liquid can always be increased by

a. increasing the pressure of the gas above the solvent.

b. increasing the temperature of the solvent.

c. decreasing the temperature of the solvent.

d. decreasing the temperature of the gas above the solvent.

e. decreasing the pressure of the gas above the solvent.

9.12 If the dissolution of an ionic solid in water in endothermic, then it can be concluded that

a. the solubility of the solid will increase with increasing temperature.

b. ΔH is negative.

c. the solution process releases heat.

d. the hydration energy exceeds the lattice energy.

e. the solubility of the solid is independent of the temperature.

9.13 When solid NH_4NO_3 is dissolved in a nearly saturated solution of NH_4NO_3, the solution becomes colder. This information indicates that if the temperature of a solution is raised,

a. ΔH will become negative.

b. ΔH will become zero.

c. the solubility of NH_4NO_3 will decrease.

d. the solubility of NH_4NO_3 will increase.

e. the solubility of NH_4NO_3 will not change.

9.14 Which of the following solutions would be expected to be the poorest conductor of an electric current at room temperature?

a. 1.0 M CH_3COOH b. 0.01 M CH_3COOH c. 0.01 M $C_{12}H_{22}O_{11}$

d. 0.01 M NaCl e. 0.01 M $MgCl_2$

76% 9.11 (a) See the equation for Henry's law in the answer to question 9.10. The solubility increases as P(g) increases.

56% 9.12 (a) $MX(s) + \text{heat} \xrightleftharpoons{\text{water}} MX(aq)$

From a consideration of LeChatelier's principle, if we exert a stress (increasing the temperature) on the left hand side of the equation, it well shift to the right.

9.13 (d) See the answer to question 9.12.

56% 9.14 (c) Sucrose is a molecule that doesn't ionize. All the other solutes in this question will produce some ions in aqueous solution.

9.15 Gaseous hydrogen chloride is added to water and methylbenzene. The electrical conductivity of HCl in water is much much greater than HCl in methylbenzene. This is due primarily to the fact that

a. water is intrinsically a conducting solvent.

b. water is a better proton-acceptor than methylbenzene.

c. HCl has a higher solubility in water than in methylbenzene.

d. water is a better proton donor than methylbenzene.

e. water and methylbenzene are immiscible.

9.16 Which of the following ions would be expected to have the largest heat of hydration?

a. Na^+ b. Al^{3+} c. S^{2-} d. Ba^{2+} e. F^-

9.17 Which of the following aqueous solutions is the best conductor of an electric current at room temperature?

a. 0.1 M sucrose b. 0.1 M acetic acid c. 0.1 M ammonium carbonate

d. 0.1 M sodium chloride e. 0.005 M sodium chloride

9.18 Which of the following solutes dissolved in 1000. g of water would provide the fewest number of particles?

a. 0.010 moles nitric acid, HNO_3

b. 0.010 moles hydrochloric acid, HCl

c. 0.010 moles nitrous acid, HNO_2

d. 0.010 moles sulfuric acid, H_2SO_4

e. 0.010 moles chloric acid, $HClO_3$

9.19 Which of the following solutes dissolved in 1000 g of water would provide the largest number of particles?

a. 0.030 moles urea, $CO(NH_2)_2$

b. 0.030 moles acetic acid, CH_3COOH

c. 0.015 moles ammonium nitrate, NH_4NO_3

d. 0.015 moles calcium sulfate, $CaSO_4$

e. 0.010 moles barium chloride, $BaCl_2$

9.15 (b) $HCl(g) + H_2O(\ell) \rightarrow H_3O^+(aq) + Cl^-(aq)$
Hydrogen chloride reacts with water to produce hydrochloric acid which is a strong electrolyte.

64% 9.16 (b) The hydration energy of an ion is proportional to the charge/radius ratio of the ion. The higher the charge and the smaller the ion the greater the hydration energy.

9.17 (c) Which solution has the most ions? Because of concentration we will consider solutions (c) and (d). We should select solution (c) because the formula of ammonium carbonate is $(NH_4)_2CO_3$ so it has 3 ions per formula unit whereas NaCl has 2 ions per formula unit.

60% 9.18 (c) Nitrous acid is a weak acid so it would provide fewer particles. All the other solutes in this question are strong acids.

40% 9.19 (b) All the solutions have 0.030 moles of particles except the solution which has 0.030 moles of CH_3COOH. Since CH_3COOH is a weak acid, there will be more than 0.030 moles of particles.

$$CH_3COOH + H_2O \rightleftarrows H_3O^+ + CH_3COO^-$$

$$0.030 - y \qquad y \qquad y$$

mol particles = 0.030 -y +y +y = 0.030 +y

Lowering of Vapor Pressure and Raoult's Law

9.20 The vapor pressure of a solution containing a nonvolatile solute is directly proportional to the

a. mole fraction of solvent.

b. mole fraction of solute.

c. molality of the solvent.

d. molality of the solute.

e. osmotic pressure of the solute.

The information below is for questons 9.21 - 9.23.
Consider that an ideal solution is formed from a mixture of the nonvolatile solute, urea ($CO(NH_2)_2$), and methanol (CH_3OH). The vapor pressure of pure methanol at 20°C is 89 mm Hg.

9.21 Calculate the mole fraction of urea that is obtained from mixing 6.0 g of urea with 32.0 g of methanol.

a. 0.091 b. 0.10 c. 0.16 d. 0.84 e. 0.91

9.22 Calculate the vapor pressure of the methanol solution in problem 9.21.

a. 8.1 mm Hg b. 8.9 mm Hg c. 14 mm Hg d. 75 mm Hg

e. 81 mm Hg

9.23 Calculate the vapor pressure-lowering for the methanol solution in problem 9.21.

a. 6.0 mm Hg b. 8.0 mm Hg c. 14 mm Hg d. 75 mm Hg

e. 81 mm Hg

40% 9.20 (a) $P_{solvent} = X_{solvent} P^{\circ}_{solvent}$

Since a solution consists of a solute and a solvent and the solute is nonvolatile, then the vapor pressure of the solution is simply the vapor pressure of the solvent. The change in vapor pressure, $\Delta P_{solvent} = X_{solute} P^{\circ}_{solvent}$ is directly proportional to the mole fraction of solute.

9.21 (a) $$X_u = \frac{n_u}{n_u + n_m} = \frac{\frac{6.0}{60.0}}{\frac{6.0}{60.0} \times \frac{32.0}{32.0}} = 0.091$$

9.22 (b) $P_{solution} = X_{solvent} P^{\circ}_{solvent}$

$$P_{solution} = (1.00 - 0.091)(89 \text{ mm Hg}) = 81 \text{ mm Hg}$$

9.23 (b) $\Delta P = P_{solvent} - P_{solution} = 89\text{-}81 = 8 \text{ mm Hg}$

The data below are for questions 9.24 and 9.25.
An ideal solution is formed from a mixture of ethanol and methanol. The vapor pressure of ethanol is 44 mm Hg and that of methanol is 89 mm Hg at 20°C.

9.24 Calculate the vapor pressure of a solution obtained by mixing 138 g ethanol (M.W. = 46 g/mole) and 64 g methanol (M.W. = 32 g/mole).
a. 44 mm Hg b. 59 mm Hg c. 62 mm Hg d. 71 mm Hg
e. 89 mm Hg

9.25 The mole fraction of the methanol vapor in problem 9.24 is
a. 0.25 b. 0.43 c. 0.50 d. 0.57 e. 0.75

The data below are for questions 9.26 and 9.27.
Consider that an ideal solution is formed from a mixture of chloroform and carbon tetrachloride. At 25°C, the vapor pressure of $CHCl_3$ is 199 torr and CCl_4 is 114 torr.

9.26 If 35% of the molecules in the solution are $CHCl_3$, the vapor pressure of the solution in torr at 25°C is
a. 114 b. 144 c. 156.5 d. 169 e. 199

9.27 The mole fraction of the $CHCl_3$ vapor in problem 9.26 is
a. 0.23 b. 0.35 c. 0.48 d. 0.52 e. 0.77

The data below are for questions 9.28 and 9.29.
Benzene, C_6H_6, and toluene, $C_6H_5CH_3$, form ideal solutions. At 35°C the vapor pressure of benzene is 160. torr, while that of toluene is 50.0 torr. 3.20 moles of benzene and 4.80 moles of toluene are placed in a closed container at 35°C and allowed to come to equilibrium.

9.28 What is the mole fraction of benzene in the liquid state?
a. 0.25 b. 0.32 c. 0.40 d. 0.48 e. 0.60

60% 9.24 (c) $$X_{C_2H_5OH} = \frac{n_{C_2H_5OH}}{n_{C_2H_5OH} + n_{CH_3OH}}$$

$$= \frac{138\ g \times \frac{1\ mol}{46\ g}}{138\ g \times \frac{1\ mol}{46\ g} + 64\ g \times \frac{1\ mol}{32\ g}} = \frac{3}{3 + 2} = 0.60$$

$$X_{CH_3OH} = 1.00 - 0.60 = 0.40$$

$$P_T = X_{C_2H_5OH}\ P°_{C_2H_5OH} + X_{CH_3OH}\ P°_{CH_3OH}$$

$$= 0.60\ (44) + 0.40\ (89) = 62\ torr$$

40% 9.25 (d) $$X_{CH_3OH} = \frac{P°_{CH_3OH}}{P_T} = \frac{0.40\ (89)}{62} = 0.57$$

9.26 (b) $$P_T = X_{CHCl_3}\ P°_{CHCl_3} + X_{CCl_4}\ P°_{CCl_4}$$

$$= 0.35\ (199) + 0.65\ (114)$$
$$= 69.7 + 74.1 = 143.8\ torr$$

9.27 (c) $$X_{CHCl_3} = \frac{P_{CHCl_3}}{P_T} = \frac{69.7}{143.8} = 0.48$$

Note that the more volatile component has become more concentrated in the vapor phase.

90% 9.28 (c) $$X_b = \frac{n_b}{n_b + n_t} = \frac{3.20}{3.20 + 4.80} = 0.40$$

9.29 What is the mole fraction of benzene in the vapor phase?

a. 0.28 b. 0.32 c. 0.52 d. 0.60 e. 0.68

Freezing Point Depression and Boiling Point Elevation

9.30 Which of the following solutes in aqueous solution would be expected to exhibit the smallest freezing-point lowering (assuming ideal behavior)?

a. 0.1 m $NaCl$ b. 0.2 m CH_3COOH c. 0.1 m $MgCl_2$

d. 0.05 m $Al_2(SO_4)_3$ e. 0.25 m NH_3

9.31 Which of the following solutes dissolved in 1.0 kg of water would be expected to boil at the highest temperature?

a. 0.0030 moles acetic acid (CH_3COOH)

b. 0.0030 moles sucrose ($C_{12}H_{22}O_{11}$)

c. 0.0015 moles $MgSO_4$

d. 0.0015 moles HCl

e. 0.0010 moles $Ca(NO_3)_2$

9.32 Which of the following solutes dissolved in 1000 g of water would be expected to boil at the lowest temperature?

a. 0.003 moles glucose ($C_6H_{12}O_6$)

b. 0.003 moles acetic acid (CH_3COOH)

c. 0.002 moles NH_4NO_3

d. 0.002 moles $MgSO_4$

e. 0.002 moles $MgCl_2$

60% 9.29 (e) $P_T = X_bP^\circ_b + X_tP^\circ_t$

$$= 0.40(160) + 0.60(50) = 64 + 30 = 94 \text{ torr}$$

$$X_b = \frac{P_b}{P_T} = \frac{64}{94} = 0.68$$

9.30 (a) Calculate the number moles of particles per kg of solvent.

$$0.1 \text{ mol NaCl} \times \frac{2 \text{ mol ions}}{1 \text{ mol NaCl}} = 0.2$$

0.2 m CH_3COOH gives more than 0.20. mol (see answer to question 9.19).

$$0.1 \text{ mol } MgCl_2 \times \frac{3 \text{ mol ions}}{1 \text{ mol } MgCl_2} = 0.30$$

$$0.05 \text{ mol } Al_2(SO_4)_3 \times \frac{5 \text{ mol ions}}{1 \text{ mol } Al_2(SO_4)_3} = 0.25$$

0.25 mol NH_3 (a weak electrolyte that reacts with water see CH_3COOH above).

9.31 (a) Which solution has the most particles? Solutions (b) to (e) have 0.0030 moles of particles. Solution (a) originally had 0.0030 moles of CH_3COOH but some of the acetic acid dissociates to give more than 0.0030 moles of particles.

9.32 (a) Which solution has the fewest number of particles? Which solute produces the fewest number of particles? We need to consider only (a) and (b). Acetic acid is a weak electrolyte so it will give more particles than glucose.

9.33 Which of the following solutes dissolved in 1.0 kg of water would be expected to cause the water to freeze at the lowest temperature?

a. 0.0015 moles glucose ($C_6H_{12}O_6$)

b. 0.0015 moles HF

c. 0.0020 moles methanol (CH_3OH)

d. 0.0015 moles H_2SO_3

e. 0.0015 moles HNO_3

9.34 In order to determine the molecular weight of a nonelectrolyte by the freezing point depression method, the minimum data required are

a. ΔT and K_f only.

b. ΔT, K_f and weight of solute.

c. ΔT, K_f, weight of solute, and weight of solvent.

d. ΔT, K_f, weight of solute, weight of solvent and identity of solvent.

e. ΔT, K_f, weight of solute, weight of solvent, identity of solvent and identity of solute.

9.35 The freezing point of a 0.025 m solution of formaldehyde (CH_2O) in water would be (K_f for water is 1.86°C/m)

a. -1.86°C b. -0.046°C c. 0.046°C d. 0.093°C e. -0.093°C

9.36 The freezing point of an aqueous solution of a nonelectrolyte is -0.23°C. The molality of this solution is ($K_f(H_2O)$ = 1.86°C/m)

a. 0.12 b. 0.25 c. 0.45 d. 0.51 e. 1.86

The data below are for questions 9.37 - 9.40.
A compound containing boron, nitrogen, and hydrogen analyzes 40.3% B, 52.2% N, and 7.5% H. A 3.562 g sample of this compound was dissolved in 50.00 g of benzene producing a solution freezing at 1.3°C. K_f for benzene is 5.12°C/molal. Pure benzene has a freezing point of 5.48°C.

9.33 (e) Which solution will have the most particles? Nitric acid is a strong electrolyte so there will be 0.0030 moles of particles. Glucose and methanol are nonelectrolytes and HF and H_2SO_3 are weak acids.

64% 9.34 (c) $\Delta T = K_f m = K_f \dfrac{\frac{wt}{MW}}{1.00 \text{ kg solvent}}$

The only unknown is molecular weight (MW).

9.35 (b) $\Delta T = K_f m = 1.86°C/m \times 0.025 = 0.046°C$

Freezing point = 0.00 - 0.046 = -0.046°C

90% 9.36 (a) $\Delta T = K_f m\ ;\ m = \dfrac{\Delta T}{K_f} = \dfrac{0.23°C}{1.86°C/m} = 0.12\ m$

9.37 Determine the empirical formula of the compound from the percentage composition. The empirical weight of the B, N, H compound is

a. 25.8 b. 26.8 c. 50.6 d. 51.6 e. 106

9.38 The molality of the benzene solution is

a. 0.071 b. 0.24 c. 0.70 d. 0.82 e. 1.07

9.39 The experimental molecular weight for the compound is

a. 67 b. 87 c. 102 d. 106 e. 297

9.40 The molecular formula for the compound is

a. B_2N_2H b. $B_4N_4H_2$ c. BNH_2 d. $B_3N_3H_6$ e. $B_4N_4H_8$

The information given below is to be used for questions 9.41 - 9.43. A compound containing sulfur, nitrogen, and hydrogen analyses 86.6% sulfur, 12.6% nitrogen, and 0.91% hydrogen. A 1.263 g sample of this compound was dissolved in 25.00 g of bromoform producing a solution freezing at 4.4°C. The solvent has a freezing point of 7.8°C. K_f for bromoform is 14.4°C/molal.

9.41 The empirical weight of the S, N, H compound is

a. 47.1 b. 79.2 c. 93.2 d. 111 e. 125

9.42 The molality of the bromoform solution is

a. 0.24 b. 0.31 c. 0.54 d. 0.78 e. 3.3

9.43 The molecular formula for the compound is

a. SNH b. S_3NH c. $S_2N_2H_2$ d. $S_3N_2H_2$ e. $S_6N_2H_2$

9.44 A solution consisting of 0.200 mole of methylbenzene ($C_6H_5CH_3$) in 500. g of nitrobenzene ($C_6H_5NO_2$) freezes at 3.2°C. Pure nitrobenzene freezes at 6.0°. The molal freezing point constant of nitrobenzene is

a. 2.8 °C $molal^{-1}$ b. 3.2 °C $molal^{-1}$ c. 5.6 °C $molal^{-1}$

d. 7.0 °C $molal^{-1}$ e. 14.0 °C $molal^{-1}$

84% 9.37 (b) $\# \text{ mol B} = 40.3 \text{ g B} \times \frac{1 \text{ mol B}}{10.8 \text{ g B}} = 3.73$ 1

$\# \text{ mol N} = 52.2 \text{ g N} \times \frac{1 \text{ mol N}}{14.0 \text{ g N}} = 3.73$ 1

$\# \text{ mol H} = 7.5 \text{ g H} \times \frac{1 \text{ mol H}}{1.01 \text{ g H}} = 7.43$ 2

BNH_2 ; empirical weight = 10.8 + 14.0 + 2(1.01) = 26.8

84% 9.38 (d) $m = \frac{\Delta T}{K_f} = \frac{5.48-1.3}{5.12} = 0.82 \text{ m}$

78% 9.39 (b) $\frac{3.562 \text{ g sample}}{0.05000 \text{ kg solvent}} = 71.24 \text{ g/kg}$

$MW = \frac{wt}{m} = \frac{71.24 \text{ g}}{0.82 \text{ mol}} = 87$

70% 9.40 (d) $\# \text{ empirical formula units} = \frac{87}{26.8} = 3.2$

$\therefore (BNH_2)_3 = B_3N_3H_6$

9.41 (d) $\# \text{ mol S} = 86.6 \text{ g S} \times \frac{1 \text{ mol S}}{32.0 \text{ g S}} = 2.71$ 3

$\# \text{ mol N} = 12.6 \text{ g N} \times \frac{1 \text{ mol N}}{14.0 \text{ g N}} = 0.90$ 1

$\# \text{ mol H} = 0.91 \text{ g H} \times \frac{1 \text{ mol H}}{1.01 \text{ g H}} = 0.90$ 1

$\therefore S_3NH$; 32 x 3 + 14 + 1 = 111

9.42 (a) $m = \frac{\Delta T}{K_f} = \frac{(7.8-4.4)°C}{14.4°C/m} = 0.236$

9.43 (e) $\# \text{ g/1.00 kg} = \frac{1.263 \text{ g}}{0.02500 \text{ kg}} = 50.52$

$MW = \frac{wt}{m} = \frac{50.52 \text{ g}}{0.236 \text{ mol}} = 214 \text{ g/mol}$

$\# \text{ empirical units} = \frac{214}{111} = 2$ (rounded)

$\therefore S_6N_2H_2$

9.44 (d) $K_f = \frac{\Delta T}{m} = \frac{(6.0-3.2)°C}{0.400 \text{ mol/kg}}$

The data below are for questions 9.45 - 9.48. A compound containing carbon, hydrogen and oxygen has 40.0% C, 6.7% H, and 53.3% oxygen. It is obtained from corn cobs by boiling with 8% sulfuric acid. A 3.025 g sample of this compound was dissolved in 25.00 g of pyridine producing a solution boiling at 117.6°C. K_b for pyridine is 2.78°C/molal. Pure pyridine has a normal boiling point of 115.5°C.

9.45 The empirical mass of the C, H, O compound is

a. 29 b. 30 c. 42 d. 44 e. 91

9.46 The molality of the pyridine solution is

a. 0.76 b. 1.1 c. 1.3 d. -1.3 e. 2.1

9.47 The experimental molecular weight for the compound is

a. 62 b. 80 c. 93 d. 110 e. 160

9.48 The molecular formula for the compound is

a. $C_3H_3O_3$ b. $C_4H_4O_2$ c. $C_4H_{10}O_4$ d. $C_5H_{10}O_5$ e. $C_6H_{12}O_6$

9.49 The freezing point of a solution prepared by dissolving 3.00 g of compound X (a nonelectrolyte) in 100 g of water is -0.279°C. The molecular weight of compound X in g/mol is

a. 100 b. 200 c. 300 d. 175 e. 95.0

9.50 For a 0.001 molar solution of $(NH_4)_2CO_3$ the van't Hoff factor, i would bc about

a. 1 b. 2 c. 3 d. 4 e. 5

9.51 For a 0.001 molar $Al(NO_3)_3$ the van't Hoff factor, i would be about

a. 2 b. 3 c. 4 d. 5 e. 6

9.52 For a 0.001 molal solution of $Al_2(SO_4)_3$, the van't Hoff i factor would probably be

a. ~ 1 b. ~ 2 c. ~ 3 d. ~ 4 e. ~ 5

88% 9.45 (b) $\# \text{ mol C} = 40.0 \text{ g C} \times \frac{1 \text{ mol C}}{12.0 \text{ g C}} = 3.33$ 1

$\# \text{ mol H} = 6.7 \text{ g H} \times \frac{1 \text{ mol H}}{1.01 \text{ g H}} = 6.63$ 2

$\# \text{ mol O} = 53.3 \text{ g O} \times \frac{1 \text{ mol O}}{16.0 \text{ g O}} = 3.33$ 1

$\therefore CH_2O$; 12 + 2 + 16 = 30

88% 9.46 (a) $m = \frac{\Delta T}{K_f} = \frac{(117.6 - 115.5)°C}{2.78°C/m} = 0.76 \text{ m}$

9.47 (e) $\frac{3.025 \text{ g sample}}{25.00 \text{ g solvent}} = \frac{121.0 \text{ g}}{1.00 \text{ kg}}$

$MW = \frac{wt}{m} = \frac{121.0 \text{ g}}{0.76} = 159$

72% 9.48 (d) $\# \text{ empirical formula units} = \frac{159}{30} = 5.3$

This is an experimental number and usually we are doing very well if we can use this technique and come within ± 5% of the molecular weight. Round to 5.

9.49 (b) $\frac{\# \text{ g solute}}{1.00 \text{ kg } H_2O} = \frac{3.00 \text{ g}}{0.100 \text{ kg}} = 30.0$

$m = \frac{\Delta T}{K_f} = \frac{2.79°C}{1.86°C/\text{molal}} = 0.150$

$M.W. = \frac{wt}{m} = \frac{30.0}{0.150} = 200 \text{ g/mol}$

80% 9.50 (c) How many ions are there in $(NH_4)_2CO_3$? $2\ NH_4^+ + 1\ CO_3^{2-}$

88% 9.51 (c) How many ions are there in $Al(NO_3)_3$? $1\ Al^{3+} + 3\ NO_3^-$

9.52 (e) How many ions are there in one formula unit of $Al_2(SO_4)_3$?

9.53 For which of the following aqueous solutions would one expect to have the largest van't Hoff i factor?

a. 0.100 m $C_6H_{12}O_6$ (glucose) b. 0.100 m NaCl c. 0.010 m NaCl

d. 0.100 m K_2SO_4 e. 0.010 m K_2SO_4

Osmotic Pressure

9.54 To determine the molecular weight of a small protein whose molecular weight is in the range of 20,000 - 60,000, it would be best to measure

a. the boiling point of the solvent and of the solution and the weight of the solute and solvent.

b. the freezing point of the solvent and of the solution and the weight of the solute and solvent.

c. the osmotic pressure of the solution and the weight of the solute and the volume of the solvent.

d. the vapor pressure of the solvent and of the solution and the weight of the solute and solvent.

e. the density of the protein and make x-ray measurements.

9.55 A 1.0 g sample of a small protein having a molecular weight of 50,000 g/mol is dissolved in 50.0 mL of water. Calculate the osmotic pressure of the solution in mm Hg at a temperature of 25°C.

a. 9.8×10^{-3} mm Hg b. 3.7 mm Hg c. 7.4 mm Hg

d. 14.9 mm Hg e. 21.2 mm Hg

9.56 Calculate the molecular weight of a small protein if a 0.20 gram sample dissolved in 100. mL of water has an osmotic pressure of 9.8 mm Hg at 25.0°C.

a. 3.5×10^3 g/mol b. 3.8×10^3 g/mol c. 4.0×10^4 g/mol

d. 4.5×10^4 g/mol e. None of these.

30% 9.53 (e) The i factor depends both on the solute and its concentration. At low concentrations, i equals the number of ions per formula unit. The value of i decreases with increasing concentration to reflect greater ionic association.

9.54 (c) Osmotic pressures can be measured for very dilute solutions (whereas freezing point depressions and boiling point elevations can only be measured for relatively concentrated solutions). The consequence of this fact means that we can determine the molecular weights of large molecules. See question 9.55.

35% 9.55 (c) $\pi = MRT$

$$\pi = \frac{\frac{1.0\ \text{g}}{50{,}000\ \text{g/mol}}}{0.0500\ \text{L}} \times 0.08205\ \frac{\text{L}\cdot\text{atm}}{\text{mol}\cdot\text{K}} \times 298\ \text{K}$$

$$= 9.78 \times 10^{-3}\ \text{atm}$$

$$\pi = 9.78 \times 10^{-3}\ \text{atm} \times 760.\ \text{mm Hg/atm} = 7.4\ \text{mm Hg}$$

9.56 (b) $$M = \frac{\pi}{RT} = \frac{9.8\ \text{mm Hg} \times \frac{1\ \text{atm}}{760.\ \text{mm Hg}}}{0.08205\ \frac{\text{L}\cdot\text{atm}}{\text{mol K}} \times 298\ \text{K}} = 5.27 \times 10^{-4}\ \text{mol/L}$$

$$\text{M.W.} = \frac{\text{wt}}{\text{mol}} = \frac{2.0\ \text{g/L}}{5.27 \times 10^{-4}\ \text{mol/L}} = 3.8 \times 10^{3}\ \text{g/mol}$$

10

ACIDS AND BASES

Acidic and Basic Character

10.1 Which of the following nonmetal oxides is paired with the acid which will result from the addition of water to the nonmetal oxide?

a. NO_2 and HNO_2 b. SO_3 and H_2SO_4 c. P_4O_{10} and H_3PO_3
d. N_2O_5 and HNO_2 e. NO and HNO_2

10.2 The acidic anhydride of H_2SO_3 is

a. SO_3 b. S_2O_3 c. SO_2 d. S_2O_6 e. S_2O_7

10.3 All the following are acid anhydrides EXCEPT

a. Tl_2O_3 b. N_2O_3 c. N_2O_5 d. P_4O_6 e. P_4O_{10}

10.4 The acid anhydride of perchloric acid is

a. Cl_2O b. Cl_2O_3 c. ClO_2 d. Cl_2O_5 e. Cl_2O_7

10.5 When Na_2O is added to water, the solution is basic because it contains and excess of

a. NaO b. Na^+ c. O_2^- d. O^{2-} e. OH^-

10.6 The solutes listed below were all added to water. Which of the solutions is the most acidic? The original concentrations of all the solutes were 0.03 M.

a. SO_3 b. SO_2 c. CO_2 d. HF e. Li_2O_2

10.7 Which one of the following salts is <u>not</u> derived from a strong acid and a strong base?

a. $BaCl_2$ b. KCl c. $NaNO_3$ d. Na_2SO_3 e. LiBr

50% 10.1 (b) $SO_3 + H_2O \rightarrow H_2SO_4$

The oxidation number of the central atom in the acidic anhydride has to be the same as that in the acid. Sulfur above has a +6 oxidation state in both SO_3 and H_2SO_4.

90% 10.2 (c) $SO_2 + H_2O \rightarrow H_2SO_3$

Sulfur has +4 oxidation number in both SO_2 and H_2SO_3.

80% 10.3 (a) Acid anhydrides are nonmetal oxides. Metal oxides (Tl_2O_3) are basic anhydrides.

90% 10.4 (e) $Cl_2O_7 + 2H_2O \rightarrow 2HClO_4$

Chlorine has a +7 oxidation number in both Cl_2O_7 and $HClO_4$.

88% 10.5 (e) $Na_2O(s) + 2H_2O \rightarrow 2Na^+(aq) + 2OH^-(aq)$

56% 10.6 (a) $SO_3 + H_2O \rightarrow H_2SO_4$

H_2SO_4 is a strong acid whereas H_2SO_3, H_2CO_3 and HF are weak acids.

75% 10.7 (d) $H_2SO_3(aq) + 2NaOH(aq) \rightarrow Na_2SO_3(aq) + 2H_2O$

H_2SO_3 is a weak acid.

10.8 Which one of the following is a strong soluble base?

a. NH_4Cl b. $Mg(OH)_2$ c. $Ba(OH)_2$ d. $Cu(OH)_2$ e. $Fe(OH)_3$

10.9 Which one of the following is the most basic oxide?

a. N_2O_3 b. N_2O_5 c. P_4O_6 d. P_4O_{10} e. Bi_2O_3

10.10 All the following are strong acids and weak acids respectively EXCEPT

a. HNO_3 and HNO_2 b. H_2SO_4 and H_2SO_3 c. $HClO_4$ and $HClO$

d. $HClO_3$ and $HClO_2$ e. H_3PO_4 and $HClO$

10.11 All the following are strong acids and weak acids respectively EXCEPT

a. HBr and HF b. H_2SO_3 and HNO_2 c. HCl and H_3PO_4

d. HNO_3 and CH_3COOH e. $HClO_4$ and $HClO$

10.12 A strong acid and weak acid respectively are

a. HF and HCl b. H_2SO_4 and H_2SO_3 c. H_3PO_3 and H_3PO_4

d. HCl and HBr e. CH_3COOH and CH_3CH_2COOH

10.13 All of the following are strong acids EXCEPT

a. HNO_3 b. $HClO_2$ c. HCl d. HBr e. H_2SO_4

10.14 All of the following compounds are correctly identified as to type of compound EXCEPT

	Compound	Type of Compound
a.	LiOH	Strong soluble base
b.	$HClO_4$	Strong soluble acid
c.	HOCl	Weak soluble acid
d.	NH_3	Weak insoluble base
e.	CH_3CH_2OH	Soluble nonelectrolyte

80% 10.8 (c) A strong base requires the presence of the hydroxide ion. $Mg(OH)_2$, $Cu(OH)_2$ and $Fe(OH)_3$ are insoluble compounds.

10.9 (e) A basic oxide is a metal plus oxygen. Bismuth is the only metal given in this question.

88% 10.10 (e) Both H_3PO_4 and H_3PO_3 are weak acids.

88% 10.11 (b) Both H_2SO_3 and HNO_2 are weak acids. It is very helpful to memorize the seven common strong acids, HCl, HBr, HI, $HClO_4$, $HClO_3$, H_2SO_4 and HNO_3.

10.12 (b)
(a) weak, strong
(b) strong, weak
(c) weak, weak
(d) strong, strong
(e) weak, weak

88% 10.13 (b) By this time you should know the seven common strong acids. Consider that all other acids are weak.

70% 10.14 (d) Ammonia is a weak base that is extremely soluble in water.

10.15 Which of the following is a weak electrolyte in aqueous solution?

a. H_2S b. H_2SO_4 c. HI   d. HNO_3 e. $HClO_3$

10.16 Which of the following is a strong electrolyte in aqueous solution?

a. NH_3 b. CH_3COOH c. CO_2 d. HNO_2 e. NH_3 + CH_3COOH

10.17 Of the following acids, the one with the weakest conjugate base is

a. H_2SO_3 b. HCl   c. CH_3COOH d. HNO_2 e. HOCl

10.18 Of the following acids, the one with the weakest conjugate base is

a. HF   b. HI   c. CH_3COOH d. HNO_2 e. HOCl

10.19 Which one of the following is a strong acid in water?

a. HOCl   b. H_3PO_4 c. H_3PO_3 d. HNO_2 e. HI

Bronsted-Lowry Theory

10.20 All of the following statements are characteristics of acids EXCEPT

a. are proton donors.   b. neutralize bases.

c. react with nonmetals to give a salt and oxygen.

d. react with bases to give a salt and water.   e. taste sour.

10.21 All the following are true statements about bases EXCEPT

a. they have a bitter taste

b. they react with salts to form weaker or more volatile acids and a new salt.

c. they have a slippery feeling.

d. they change the colors of many indicators.

e. they react with acids to form salts and water.

10.15 (a) Strong acids are strong electrolytes and weak acids are weak electrolytes.

10.16 (e) All soluble salts form strong electrolytes. An acid plus a base yields a salt.

10.17 (b) This question is really asking which of the acids is a strong acid. Strong acids have weak conjugate bases.

10.18 (b) I^- is a weak conjugate base (in other words HI is a strong acid).

10.19 (e) The seven common strong acids are HNO_3, H_2SO_4, $HClO_4$, $HClO_3$, HCl, HBr and HI (HF is a weak acid).

90% 10.20 (c) Nonmetal oxides react with water to produce acids.

92% 10.21 (b) Acids react with salts to form weaker or more volatile acids and a new salt.

10.22 In the equation, $HCN(aq) + H_2O \rightleftharpoons H_3O^+(aq) + CN^-(aq)$

a. H_3O^+ is an acid and H_2O is its conjugate base.
b. HCN is an acid and CN^- is its conjugate base.
c. HCN is an acid and H_2O is its conjugate base.
d. HCN is a base and H_2O is its conjugate acid.
e. Both HCN and CN^- can function as acids.

10.23 Which of the following can be shown to be both an acid and a base according to the Bronsted-Lowry definitions?

a. H_3O^+ b. CO_3^{2-} c. HCO_3^- d. NH_4^+ e. BF_3

10.24 Which of the following would not be expected to function as a Bronsted-Lowry base?

a. NH_3 b. NH_4^+ c. O^{2-} d. H_2O e. OH^-

10.25 In the reversible reaction, $HCO_3^-(aq) + OH^-(aq) \rightleftharpoons CO_3^{2-}(aq) + H_2O$, the Bronsted acids are

a. HCO_3^- and CO_3^{2-} b. HCO_3^- and H_2O c. OH^- and H_2O

d. OH^- and CO_3^{2-} e. HCO_3^-, H_2O and OH^-

10.26 In which reaction is $H_2PO_4^-$ acting as a base?

a. $H_2PO_4^- + CN^- \rightleftharpoons HCN + HPO_4^{2-}$
b. $H_2PO_4^- + OH^- \rightleftharpoons H_2O + HPO_4^{2-}$
c. $H_2PO_4^- + HS^- \rightleftharpoons H_2S + HPO_4^{2-}$
d. $H_2PO_4^- + HF \rightleftharpoons F^- + H_3PO_4$
e. $H_2PO_4^- + NH_3 \rightleftharpoons NH_4^+ + HPO_4^{2-}$

70% 10.22 (b) An acid-conjugate base pair differ by a proton. The reactants represent the acid and base and the products represent the conjugate acid and conjugate base.

10.23 (c) $HCO_3^- + H_2O \rightleftarrows H_3O^+ + CO_3^{2-}$ (an acid)

$HCO_3^- + H_2O \rightleftarrows H_2CO_3 + OH^-$ (a base)

or $HCO_3^- + H^+ \rightleftarrows H_2CO_3$ (a base)

10.24 (b) NH_4^+ is not capable of accepting a proton as all the valence orbitals on nitrogen are filled.

10.25 (b) Both HCO_3^- and H_2O are proton donors.

10.26 (d) $H_2PO_4^-$ accepts a proton from HF so it is a base.

10.27 Which of the following would not be expected to function as a Bronsted-Lowry base?

a. SO_4^{2-} b. HSO_4^- c. NH_3 d. N_2H_4 e. C_2H_6

10.28 All of the following are acid-base conjugate pairs EXCEPT

a. H_3O^+, OH^- b. H_2O, OH^- c. NH_4^+, NH_3 d. CH_3COOH, CH_3COO^-

e. HPO_4^{2-}, PO_4^{3-}

10.29 All of the following are acid-base conjugate pairs EXCEPT

a. NH_3 and NH_2^- b. OH^- and O^{2-} c. NH_4^+ and NH_3

d. HCN and HOCN e. H_3PO_4 and $H_2PO_4^-$

10.30 $HF + H_2O \rightleftarrows H_3O^+ + F^-$

The conjugate base of HF is

a. HF b. H_2F^+ c. F^- d. H_2O e. OH^-

10.31 The conjugate base of HNO_2 is

a. HNO_2 b. H_3O^+ c. H^+ d. NO_2^- e. H_2O

Lewis Acids and Bases

10.32 $2Na^+ + 2OH^- + H^+ + HSO_4^- \rightarrow 2Na^+ + SO_4^{2-} + 2H_2O$

The acid-base reaction above can be explained by

a. only the Arrhenius theory.

b. only the Bronsted-Lowry theory.

c. only the Lewis theory.

d. both the Arrhenius and Bronsted-Lowry theories.

e. the Arrhenius, Bronsted-Lowry and Lewis theories.

10.27 (e) A base is a proton acceptor. A base needs a valence orbital with a lone pair of electrons so that it can accept a proton. The carbon atom has four bonds and an octet of electrons so it cannot form additional bonds.

88% 10.28 (a) An acid-base conjugate pair differs by H^+. Addition of H^+ or removal of H^+ gives the other member of the pair.

80% 10.29 (d) HCN and HOCN are two different acids.

10.30 (c) The conjugate base plus H^+ yields the parent acid.

10.31 (d) $H^+ + NO_2^- \rightarrow HNO_2$, so NO_2^- is the conjugate base.

10.32 (e) The reaction is an Arrhenius acid-base reaction because the acid consists of protons and the base consists of hydroxide ions. It is also a Bronsted-Lowry reaction because the acid is a proton donor and the base is a proton acceptor. It is also a Lewis acid-base reaction since H^+ is an electron pair acceptor and OH^- is a Lewis base because it is an electron pair donor.

10.33 On the basis of its structure, which of the following can function neither as a Lewis acid nor as a Lewis base?

a. NH_4^+ b. NH_2^- c. NH_3 d. BF_3 e. H_2O

10.34 All of the following species would be expected to function as Lewis bases EXCEPT

a. AlH_3 b. CH_3^- c. NH_3 d. H_2O e. H^-

10.35 All of the following would be expected to serve as a Lewis acid EXCEPT

a. $AlCl_3$ b. NO^+ c. Fe^{3+} d. SbF_5 e. NH_3

10.36 All of the following are Lewis acids EXCEPT

a. BCl_3 b. CCl_4 c. $SnCl_4$ d. H^+ e. Fe^{3+}

10.37 All of the following species would be expected to be able to function as Lewis bases EXCEPT

a. OH^- b. H_2O c. CH_3^+ d. NH_3 e. PF_3

10.38 All of the following species would be expected to function as Lewis bases EXCEPT

a. NH_2^- b. NH^{2-} c. OH^- d. CH_3OCH_3 e. CH_4

10.39 All of the following would be expected to be able to function as Lewis acids EXCEPT

a. BF_3 b. CH_3^+ c. NF_3 d. $BeCl_2$ e. Al^{3+}

10.33 (a) The ammonium ion has neither an empty valence orbital nor a valence orbital with a lone pair of electrons in it.

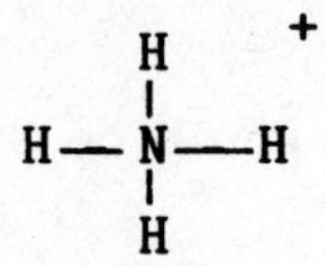

10.34 (a) All the species except AlH_3 have a lone pair of electrons about the central atom. Al has three bonding pairs so it can accept another pair of electdrons.

10.35 (e) NH_3 cannot accept a pair of electrons. It is a Lewis base because it has available a lone pair of electrons.

10.36 (b) The central carbon atom in CCl_4 has no empty valence orbitals.

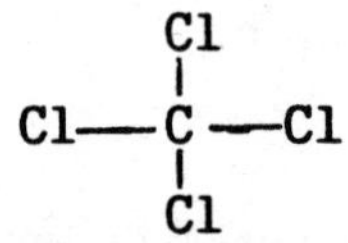

90% 10.37 (c) A Lewis base is an electron pair donor.

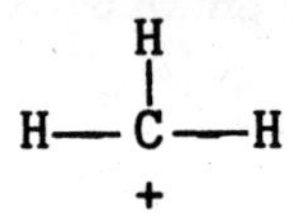

does not have a lone pair of electrons.

10.38 (e) All of the carbon valence orbitals are filled in CH_4. Methane is a saturated hydrocarbon.

60% 10.39 (c) NF_3 is a Lewis base because the central nitrogen atom has a lone pair valence electrons.

11

KINETICS

Reaction Rates and Rate Constants

11.1 The rate of a chemical reaction usually varies with

I. concentration   II. time   III. temperature   IV. catalyst

a. I only   b. I and III only   c. III and IV only

d. I, III and IV   e. I, II, III and IV

11.2 For the reaction, $I + I \rightarrow I_2$, at 25°C in CCl_4 $k = 8.2 \times 10^9$ L $mol^{-1}s^{-1}$. This reaction is

a. first order.   b. second order.   c. endothermic.

d. can't tell without knowing the activation energy.

e. can't tell without an experiment showing the concentration dependence of the iodine atom.

11.3 For the reaction, $6CH_2O + 4NH_3 \rightarrow (CH_2)_6N_4 + 6H_2O$, the rate is expressed as

$\frac{1}{6}\frac{\Delta[H_2O]}{\Delta t}$. An equivalent would be

a. $-\frac{\Delta[(CH_2)_6N_4]}{\Delta t}$ b. $6\frac{\Delta[CH_2O]}{\Delta t}$ c. $-6\frac{\Delta[CH_2O]}{\Delta t}$

d. $-\frac{1}{4}\frac{\Delta[NH_3]}{\Delta t}$ e. $-\frac{1}{6}\frac{\Delta[H_2O]}{\Delta t}$

52% 11.1 (e) The rate of a chemical reaction varies with concentration, temperature and a catalyst. It also varies with time because the concentrations of the reactants are changing with time.

11.2 (b) We cannot tell by looking at a chemical equation what the reaction order is. We can tell what the order of a reaction is from the units of the rate constant.

Rate = k $[I]^2$

Assigning units to the rate, k and [I], we get

$$\frac{mol}{L\cdot s} = \frac{L}{mol\cdot s}\left(\frac{mol}{L}\right)^n$$

Rearranging the above equation, we get

$$\frac{mol^2}{L^2} = \left(\frac{mol}{L}\right)^n$$

It can now be seen that n must be equal to two.

11.3 (d) The rate of the reaction is the change in concentration divided by the change in time ($\Delta c/\Delta t$ or dc/dt) divided by the appropriate stoichiometric coefficient and converted to a positive number.

11.4 The hypochlorite ion oxidizes the iodide ion in aqueous solution as represented by the equation, $OCl^- + I^- \rightarrow OI^- + Cl^-$. The rate of formation of OI^- is given by the rate law

$$\text{Rate} = k \frac{[I^-][OCl^-]}{[OH^-]}$$

The overall reaction order is

a. 0 b. 1 c. 2 d. 3 e. 4

11.5 The rate law for the chemical reaction, $5Br^- + BrO_3^- + 6H^+ \rightarrow 3Br_2 + 3H_2O$ has been determined experimentally to be

$$\frac{-d[BrO_3^-]}{dt} = k[Br^-][BrO_3^-][H^+]^2$$

The reaction order with respect to the hydrogen ion is

a. 1 b. 2 c. 6 d. $k[H^+]^2$ e. $k[Br^-][BrO_3^-][H^+]^2$

11.6 Nitric oxide reacts with hydrogen at a measurable rate at 1000 K according to the equation, $2NO + 2H_2 \rightarrow N_2 + 2H_2O$. The experimental rate law is Rate = $k[NO]^2[H_2]$. When time is in minutes and the concentration is in moles/liter, the units for k are

a. $\frac{\text{moles}}{\ell\text{-min}}$ b. $\frac{\text{-moles}}{\ell\text{-min}}$ c. $\frac{\ell^2}{\text{moles}^2\text{-min}}$ d. $\frac{\text{moles}^2}{\ell^2\text{-min}}$

e. $\frac{\text{moles}^3}{\ell^3\text{-min}}$

11.4 (b) $\text{Rate} = k \dfrac{[I^-]^1[OCl^-]^1}{[OH^-]}$

$\text{Order} = \dfrac{1 + 1}{1} = 1$

11.5 (b) Since the concentration dependence of H^+ is to the second power, the reaction order is to the second power.

11.6 (c) Substitute the appropriate units into the rate law and solve for k.

$$\frac{\text{mol}}{\ell\cdot\text{min}} = k\left(\frac{\text{mol}}{\ell}\right)^2\left(\frac{\text{mol}}{\ell}\right)$$

$$k = \frac{\ell^2}{\text{mol}^2\cdot\text{min}}$$

11.7 For a certain reaction, the rate expression is given by the equation below

$$\text{Rate} = k\,[X]^{3/2}[Y]^{1/2}$$

The rate has the units $\frac{\text{mole}}{\text{L}\cdot\text{s}}$.

Which of the following are possible units for the rate constant, k?

a. $\frac{\text{mole}}{\text{L}\cdot\text{s}}$ b. $\frac{1}{\text{s}}$ c. $\frac{\text{L}^2}{\text{mol}^2\cdot\text{s}}$ d. $\frac{\text{L}}{\text{mol}\cdot\text{s}}$ e. $\frac{\text{mol}^2}{\text{L}^2}$

11.8 For the net reaction, $NH_4^+ + HNO_2 \rightarrow N_2 + 2H_2O + H^+$, the rate law for the formation of H^+ is

a. $R = k\,[NH_4^+][HNO_2]$ b. $R = k\,\frac{[NH_4^+][HNO_2]}{[N_2][H_2O]^2[H^+]}$ c. $R = k\,[H^+]$

d. $R = k\,\frac{[HNO_2]}{[H^+]}$ e. Cannot tell from the data above.

11.9 When 100 cm^3 of 1.0 M HCl is added to 2 g of Na_2CO_3 crystals, carbon dioxide is evolved. All of the following will increase the initial rate of the reaction EXCEPT

a. addition of 100 cm^3 of 2.0 M HCl instead of 1.0 M HCl.

b. addition of 200 cm^3 of 1.0 M HCl instead of 100 cm^3.

c. addition of 100 cm^3 of 1.0 M HCl to powdered Na_2CO_3 instead of crystalline Na_2CO_3.

d. addition of 100 cm^3 of hot 1.0 M HCl instead of 1.0 M HCl at room temperature.

e. addition of 200 cm^3 of 2.0 M HCl instead of 100 cm^3 of 1.0 M HCl.

72% 11.7 (d) Substitute the appropriate units into the rate law and solve for k.

$$\frac{mol}{L \cdot s} = k(\frac{mol}{L})^{3/2}(\frac{mol}{L})^{1/2}$$

32% 11.8 (a) The rate law must be determined experimentally. It cannot be obtained from a consideration of the stoichiometric equation.

11.9 (b) The concentration of HCl is the same in option (b) as it is in the original problem so the addition of more HCl which is already in excess will not increase the rate.

11.10 The gas-phase decomposition of N_2O_5 is a first order process with a rate constant of $1.50 \times 10^{-3}s^{-1}$ at 55°C. The decomposition reaction is $N_2O_5(g) \rightarrow 2NO_2(g) + 1/2\ O_2(g)$. Given the following data: 5.0 g of N_2O_5 is placed in the first vessel and 2.5 g of N_2O_5 in the second vessel. The vessels are at the same temperature and pressure. How much time is required for 1/2 of the N_2O_5 to decompose in each vessel?

a. Vessel one requires twice as much time as vessel two.

b. Vessel two requires twice as much time as vessel one.

c. Vessel one requires three times as much time as vessel two.

d. Vessel one requires four times as much time as vessel one.

e. Vessel one requires the same amount of time as vessel two.

11.11 The rate law for the hydrolysis of thioacetamide

$$CH_3 - \overset{\overset{\displaystyle S}{\|}}{C} - NH_2 + H_2O \rightarrow H_2S + CH_3 - \overset{\overset{\displaystyle O}{\|}}{C} - NH_2$$

is rate = $k[H^+][CH_3CSNH_2]$. If some solid sodium hydroxide is added to a solution that is 0.10 M in both thioacetamide and hydrogen ion at 25°C then

a. the reaction rate decreases, but k remains the same.

b. the reaction rate increases, but k remains the same.

c. the reaction rate remains the same but k decreases.

d. the reaction rate remains the same but k increases.

e. There is no change in the reaction rate nor the rate constant.

11.12 If $\Delta H = +125$ kJ for a certain process, then

a. it occurs rapidly.

b. it does not need a catalyst.

c. it does need a catalyst.

d. it is exothermic.

c. it is endothermic.

11.10 (e) For first order reactions, the time required for half of the reactant to react is a constant.

$$t_{1/2} = \frac{0.693}{k}$$

The time for half of a reactant to react is called a half-life.

25% 11.11 (a) This is a difficult question for most students since OH^- does not appear directly in the rate law. This prompted 55% of the students to select option (e). The sodium hydroxide decreases the H^+ concentration.

88% 11.12 (e) The change in enthalpy for a reaction does not tell us anything about the rate of a chemical reaction. The sign of ΔH tells us whether the reaction is endothermic or exothermic.

11.13 The isotope, $^{178}_{71}Lu$ undergoes beta decay and after 2.5 hours, 1/32 of the initial amount of $^{178}_{71}Lu$ remains undecayed. The half-life of the isotope is

a. 15 min. b. 20 min. c. 30 min. d. 45 min. e. 60 min.

11.14 ^{64}Co decays by a first order process by the emission of a beta particle. The ^{64}Co isotope has a half-life of 7.8 min. How long will it take for 15/16 of the cobalt to undergo decay?

a. 7.8 min. b. 15.6 min. c. 23.4 min. d. 31.2 min. e. 39.0 min.

11.15 If a reaction is first order with a rate constant of $5.48 \times 10^{-2} sec^{-1}$, how long will it take for 3/4 of the initial concentration of the reactant to be consumed?

a. 36.5 sec b. 18.2 sec c. 25.3 sec d. 12.6 sec e. 6.3 sec

11.16 Which of the following corresponds to the correct integrated expression for a reaction involving one reactant which is second-order?

a. $\frac{1}{A} - \frac{1}{B} = kt$ b. $\ln \frac{A_o}{A} = kt$ c. $\log \frac{A_o}{A} = kt$

d. $\frac{1}{A} - \frac{1}{A_o} = kt$ e. $\frac{1}{A_o} - \frac{1}{A} = kt$

11.17 A reaction which is second order has a rate constant of 1.0×10^{-2} L $mol^{-1}s^{-1}$. If the initial concentration of the reactant is 0.100 M, how long will it take for the concentration to become 0.0500 M?

a. 100 s b. 500 s c. 1000 s d. 1500 s e. 10,000 s

11.13 (c)

1	
	1
1/2	
	2
1/4	
	3
1/8	
	4
1/16	
	5
1/32	

By decreasing the "original" amount by half for each half-life, we can see that there are five half-lifes.

$$\therefore \frac{2.5 \text{ hr} \times 60 \text{ min/hr}}{5 \text{ half-lifes}} = \frac{30 \text{ min}}{\text{half-life}}$$

11.14 (d) From a consideration of the solution to question 11.13, we can see that four half-lifes are required to reach 1/16 of the cobalt-64. $\therefore$ 4 half-lifes x $\frac{7.8 \text{ min}}{\text{half-life}}$ = 31.2 min. Alternatively we could solve for k

$$k = \frac{0.693}{t\ 1/2} = \frac{0.693}{7.8 \text{ min}} = 0.0888 \text{ min}^{-1}$$

Now substitute into the integrated form of the first order law. $-\ln\frac{c}{c_o} = kt$; $t = \frac{-\ln c/_{co}}{k} = \frac{-\ln \frac{1/16}{1}}{0.0888 \text{ min}^{-1}} = 31.2 \text{ min}$

44% 11.15 (c) $-\ln \frac{c}{c_o} = kt$; $t = \frac{-\ln c/co}{k} = \frac{-\ln 1/4\ /1}{5.48 \times 10^{-2}\ s^{-1}} = 25.3$ s

11.16 (d) $2A \rightarrow P$; Rate $= \frac{-d[A]}{dt} = k\ [A]^2$; Rearranging $\frac{-d[A]}{[A]^2} = kdt$

$$-\int_{A_o}^{A} \frac{dA}{A^2} = k \int_o^t dt \ ; \ \frac{1}{A} \Big|_{A_o}^{A} = kt \Big|_o^t \ ; \ \frac{1}{A} - \frac{1}{A_o} = kt$$

11.17 (c) $t = \frac{\frac{1}{A} - \frac{1}{A_o}}{k} = \frac{\frac{1}{0.0500} - \frac{1}{0.100}}{1.0 \times 10^{-2}} = 1000$ s

Reaction Orders and Rate Laws

11.18 For the reaction, $(CH_3)_3CCl + OH^- \rightarrow (CH_3)_3COH + Cl^-$, it is experimentally found that doubling the concentration of $(CH_3)_3CCl$ causes the reaction rate to be increased two-fold, but doubling the concentration of OH^- has no effect on the rate. The rate equation is

a. $R = k\,[(CH_3)_3CCl][OH^-]$ b. $R = k\,[(CH_3)_3CCl]^2[OH^-]$

c. $R = k\,[(CH_3)_3CCl]$ d. $R = k\,[(CH_3)_3COH][Cl^-]$

e. $R = k\,\dfrac{[(CH_3)_3COH][Cl^-]}{[(CH_3)_3CCl][OH^-]}$

11.19 For the reaction between nitrogen(II) oxide and chlorine to produce nitrosyl chloride, $2\ NO + Cl_2 \rightarrow 2\ NOCl$, it is found that tripling the concentration of both reactants increases the rate by a factor of 27. If only the concentration of chlorine is tripled, the rate increases by a factor of 3. The order of this reaction with respect to NO is

a. 0   b. 1/2   c. 1   d. 2   e. 3

The information below is to be used for questions 11.20 - 11.22. The stoichiometric equation for the reaction of nitric oxide with oxygen is $2\ NO + O_2 \rightarrow 2\ NO_2$. The data below were obtained at 273°C.

Experiment	Initial Molar Concentration (mole/L) NO	O_2	Initial Rate Mole/L•sec
I	0.020	0.010	0.032
II	0.080	0.020	1.01
III	0.040	0.040	0.50
IV	0.040	0.020	0.25
V	0.030	0.010	0.71

11.18 (c) The rate of reaction is directly proportional to the concentration of $(CH_3)_3CCl$ and the rate is independent of the concentration of OH^-. Therefore, $[OH^-]$ is to the zero power.

50% 11.19 (d) $\frac{R_2}{R_1} = \frac{[NO]^m[Cl_2]^n}{[NO]^m[Cl_2]^n}$

$$\frac{3}{1} = \frac{[3]^n}{[1]^n} \quad \therefore \quad n = 1$$

$$\frac{27}{1} = \frac{[3]^m[3]^1}{[1]^m[1]^1}$$

$$(3)^m = 9 \quad ; \quad m = 2$$

11.20 Doubling the concentration of oxygen while holding the concentration of NO constant increases the rate of reaction by a factor of

a. 1 b. 2 c. 3 d. 4 e. 8

11.21 Quadrupling the concentration of NO while holding the concentration of O_2 constant increases the rate of reaction by a factor of

a. 2 b. 4 c. 8 d. 16 e. 32

11.22 The rate disappearance of oxygen is equal to

a. $k\,[NO][O_2]$ b. $k\,\frac{[NO][O_2]}{[NO_2]}$ c. $k\,\frac{[NO]^2[O_2]}{[NO_2]}$ d. $k\,[NO][O_2]^2$

e. $k\,[NO]^2[O_2]$

11.23 Two reactants A and B are mixed and the reaction is timed until a color change occurs. The data are:

[A]	[B]	Time, Sec.
0.100	0.140	25
0.050	0.140	50
0.100	0.070	100

The order of the reaction in terms of [A] is

a. 0 b. 1/2 c. 1 d. 3/2 e. 2

The data below are to be used for questions 11.24 - 11.27. The reaction between selenious acid and iodide ion in acid solution is

$$H_2Se_2O_3 + 6\,I^- + 4H^+ \rightarrow Se(s) + 2I_3^- + 3H_2O$$

11.20 (b) Compare Experiments IV and III. When the concentration of O_2 is doubled, the rate is doubled. The rate of the reaction is therefore first order in oxygen.

11.21 (d) Compare Experiments IV and II. When the concentration of NO is doubled and the concentration of O_2 is held constant, the rate is quadrupled. Therefore, the rate dependence of NO is to the second power. Therefore, Rate is $\alpha[NO]^2$ and $(\frac{4}{1})^2 = 16$.

11.22 (e) The reaction is second order in NO and first order in O_2.

68% 11.23 (c) A time is given and time is not a rate. The rate is proportional to 1/time, so the more rapid the rate the shorter the time. Consider the first two sets of data (when the concentration of B is the same). When the concentration of A is doubled, the time is halved so the reaction must be first order with respect to A.

The data below were measured at 0°C.

Experiment	Initial Molar Concentrations			Initial Rate of Disappearance of $H_2Se_2O_3$, mole/ℓ-sec
	$[H_2Se_2O_3]$	$[H^+]$	$[I^-]$	
I	1.00×10^{-4}	2.00×10^{-2}	3.00×10^{-2}	5.30×10^{-7}
II	2.00×10^{-4}	2.00×10^{-2}	3.00×10^{-2}	1.06×10^{-6}
III	3.00×10^{-4}	4.00×10^{-2}	3.00×10^{-2}	6.36×10^{-6}
IV	3.00×10^{-4}	8.00×10^{-2}	3.00×10^{-2}	2.54×10^{-5}
V	3.00×10^{-4}	8.00×10^{-2}	6.00×10^{-2}	2.04×10^{-4}
VI	2.00×10^{-4}	2.00×10^{-2}	6.00×10^{-2}	8.48×10^{-6}

11.24 Quadrupling the concentration $H_2Se_2O_3$ while holding the concentration of H^+ and I^- constant increases the rate of the reaction by a factor of

a. 1 b. 2 c. 3 d. 4 e. 8

11.25 Doubling the concentration H^+ while holding the concentrations of $H_2Se_2O_3$ and I^- constant increases the rate of the reaction by a factor of

a. 2 b. 4 c. 6 d. 16 e. 32

11.26 Tripling the concentration of I^- while holding the concentrations of $H_2Se_2O_3$ and H^+ constant increases the rate of the reaction by a factor of

a. 3 b. 6 c. 9 d. 12 e. 27

11.27 The rate of disappearance of $H_2Se_2O_3$ is equal to

a. $K[H_2Se_2O_3][H^+][I^-]$ b. $k[H_2Se_2O_3][H^+][I^-]^2$

c. $k[H_2Se_2O_3][H^+]^2[I^-]^2$ d. $k[H_2Se_2O_3][H^+]^2[I^-]^3$

e. $k[H_2Se_2O_3]^2[H^+]^2[I^-]^3$

11.24 (d) Compare Experiments I and II. Doubling the concentration of $H_2S_2O_3$ results in the rate doubling. Therefore, in the rate law $H_2S_2O_3$ is first order.

11.25 (b) Compare Experiments III and IV. Doubling the concentration of H^+ results in the rate quadrupling. Therefore, the reaction is second order in H^+. $(2/1)^2 = 4/1$.

11.26 (e) Compare Experiments IV and V. Doubling the concentration I^- results in the rate increasing by a factor of 8. Therefore, I^- appears in the reaction to the third power. $(3/1)^3 - 27/1$

11.27 (d)

Rates, Temperature, and Activation Energy

11.28 Which of the following questions are answered by kinetic principles instead of thermodynamic principles?

I. How fast will a reaction be?

II. When the reactants are combined, will the reaction be spontaneous?

III. If a reaction occurs, what are the energy changes?

IV. What are the steps or the mechanism in the reaction?

a. I and II b. II and III c. I and IV d. II and IV e. III and IV

11.29 All of the following would increase the rate of a given chemical reaction EXCEPT

a. a large negative ΔG value.

b. size of solid reactant particles.

c. reaction temperature.

d. concentration of the reactants.

e. a catalyst.

11.30 The speed of a chemical reaction

a. is constant at any given temperature, other factors being equal.

b. is independent of the amount of contact surface of a solid involved.

c. between gases should in all cases be extremely rapid because the average kinetic energy of the molecules is great.

d. is approximately doubled for a 10°C rise in temperature.

e. varies inversely with the absolute temperature.

11.31 Which of the following statements is always TRUE.

a. Exothermic reactions have lower activation energies than endothermic reactions.

b. The rate for a reaction depends on the concentrations of all the reactants.

c. The rate of a catalyzed reaction is dependent upon the concentration of the catalyst.

d. The specific rate constant is dependent upon the concentrations of the reacting species.

e. The rate law can be determined from the stoichiometric equation.

75% 11.28 (c) Thermodynamics is concerned with the energy changes in a reaction. Kinetics is concerned with the reaction mechanism, the rate of reaction and the temperature dependence of the reaction.

11.29 (a) ΔG, the Gibbs free energy gives information on the feasibility of the reaction not on the rate of the reaction.

11.30 (d)
(a) The rate of a reaction decreases with time because the concentration of the reactants is decreasing.
(b) is dependent of the amount of contact surface of a solid involved.
(c) The rate depends upon the magnitude of the activation energy.
(d) For many chemical reactions with 10°C rise in temperature, the number of molecules with enough energy to react approximately doubles.
(e) The rate varies proportionately with the absolute temperature.

11.31 (c)
(a) Each reaction has its own characteristic activation energy.
(b) A reaction can be zero order in a reactant. See question 11.18.
(d) The value of a specific rate constant is when the concentrations of the reacting species are all 1.00 molar.
(e) No, see question 11.8.

11.32 The activation energy of a reaction can be decreased by

a. raising the temperature.

b. lowering the temperature.

c. adding more reactants.

d. removing some products.

e. adding a catalyst.

11.33 A catalyst speeds up a chemical reaction by

a. increasing the activation energy.

b. providing an alternate reaction pathway.

c. shifting the equilibrium.

d. increasing the enthalpy for the reaction.

e. decreasing the enthalpy for the reaction.

11.34 The activation energy for a reaction can be determined from the slope of the graph that comes from a plot of

a. k vs. T b. ln k vs. T c. log k vs. T d. log k vs. 1/T

e. k vs. 1/T

11.35 For the decomposition of one mole of nitrosyl chloride, $\Delta H = 38$ kJ

$$NOCl(g) \rightarrow NO(g) + 1/2\ Cl_2(g)$$

The activation energy for this reaction is 100. kJ. The activation energy for the reverse reaction is

a. 38 kJ b. 62 kJ c. 76 kJ d. 100 kJ e. 138 kJ

11.36 For the reaction of nitric oxide with ozone, $\Delta H = -200.$ kJ

$$NO(g) + O_3(g) \rightarrow NO_2(g) + O_2(g)$$

The activation energy for this reaction is 10. kJ. The activation energy for the reverse reaction is

a. 10 kJ b. 190 kJ c. 200 kJ d. 210 kJ e. 250 kJ

The information below is for questions 11.37 and 11.38.

For the reaction, $N_2O_4(g) \rightarrow 2NO_2(g)$, at 30°C $k = 5.1 \times 10^6\ s^{-1}$ and the activation energy is 54.0 kJ mol^{-1}.

68% 11.32 (e) The purpose of a catalyst is to provide another pathway for a reaction. This means that there is a different barrier and that the activation energy would be different.

11.33 (b) This question is essentially the same as question 11.32 except that it is expressed differently.

88% 11.34 (d) The Arrhenius equation is $k = Ae^{-E_a/RT}$

or in logarithmic form - $\ln k = \ln A - \frac{E_a}{RT}$

The latter equation is of the form of a linear equation, $y = b + mx$. When ln k is plotted against 1/T, there will be a straight line.

11.35 (b) Draw a potential energy diagram for this reaction or use the relationship.

$$\Delta H_{reaction} = E_{a_f} - E_{a_r}$$

$$38\ kJ = 100\ kJ - E_{a_r}$$

$$E_{a_r} = 100 - 38 = 62\ kJ$$

11.36 (d) $\Delta H_{reaction} = E_{a_f} - E_{a_r}$

$$-\ 200\ kJ = 10\ kJ - E_{a_r}$$

$$E_{a_r} = 210\ kJ$$

11.37 This reaction is

a. first order b. second order c. exothermic

d. can't tell from the data given.

e. can't tell without an experiment showing the concentration dependence of N_2O_4.

11.38 The rate constant at 45°C is

a. $5.1 \times 10^6\ s^{-1}$ b. $1.4 \times 10^7\ s^{-1}$ c. $5.2 \times 10^7\ s^{-1}$

d. $8.3 \times 10^8\ s^{-1}$ e. $1.1 \times 10^{38}\ s^{-1}$

Reaction Mechanism

11.39 All of the following statements are true EXCEPT

a. In a series of stepwise reactions, the rate-determining step is the fast one.

b. The specific rate constant for a reaction can be changed by changing the temperature.

c. The rates of most chemical reactions change with time.

d. The specific rate constant is independent of the reactant concentrations.

e. The rate of a catalyzed reaction is always dependent on the concentration of the catalyst.

The information below is for questions 11.40 and 11.41.
The complete mechanism for a reaction is considered to occur in two steps, one of which is slow and the other fast:

$$A + B \rightarrow C + D \quad \text{(slow)}$$
$$A + C \rightarrow E + F \quad \text{(fast)}$$

11.40 The rate expression for the rate of the overall reaction is equal to

a. k[A][B] b. k[A][B][C] c. $k[A]^2[B][C]$ d. k[A][C]

e. $k[A]^2[B]$

11.37 (a) We can tell from the units of the rate constant that the reaction is first order. See the solution to question 11.2.

11.38 (b) $\ln \frac{k_2}{k_1} = \frac{E_a}{R} (\frac{T_2 - T_1}{T_1 T_2})$

$$\ln k_2 = \frac{54000 \text{ J/mol}}{8.314 \text{ J/mol}\cdot\text{K}} (\frac{318-303}{303 x 318}) + \ln 5.1 \times 10^6$$

$$\ln k_2 = 1.01 + 15.44 = 16.45$$

$$k_2 = 1.4 \times 10^7 \text{ s}^{-1}$$

80% 11.39 (a) The rate-determining step is the slowest step (or slowest reaction in the reaction sequence).

11.40 (a) In each elementary reaction, the stoichiometric coefficients are the respective exponents in the rate law for that step.

11.41 The stoichiometric equation for the reaction above is

a. $A + B \rightarrow C + D$ b. $A + B \rightarrow D + E + F$ c. $A + B + C \rightarrow D + E + F$

d. $A + C \rightarrow E + F$ e. $2A + B \rightarrow D + E + F$

11.42 The reaction, $CHCl_3(g) + Cl_2(g) \rightarrow CCl_4(g) + HCl(g)$, has been proposed to occur by the mechanism

$$Cl_2 \underset{k_{-1}}{\overset{k_1}{\rightleftharpoons}} 2\,Cl \qquad \text{rapid equilibrium}$$

$$CHCl_3 + Cl \xrightarrow{k_2} HCl + CCl_3 \qquad \text{slow}$$

$$CCl_3 + Cl \xrightarrow{k_3} CCl_4 \qquad \text{very rapid}$$

The overall experimental rate law which is consistent with this mechanism is

a. $\text{Rate} = k_1\,[Cl_2]$ b. $\text{Rate} = k_2\,[CHCl_3][Cl]$

c. $\text{Rate} = k_3\,[CCl_3][Cl]$ d. $\text{Rate} = k_{exp}\,[CHCl_3][Cl_2]$

e. $\text{Rate} = k_2K_{eq}\,[CHCl_3][Cl_2]^{1/2}$

The following information may be useful for questions 11.43 and 11.44. A suggested mechanism for the decomposition of ozone is:

$$O_3 \xrightarrow{K_{eq}} O_2 + O \qquad \text{(rapid equilibrium)}$$

$$O + O_3 \xrightarrow{k_2} 2\,O_2 \qquad \text{(slow step)}$$

11.41 (e) The stoichiometric equation is given by the sum of the elementary reactions.

11.42 (e) (1) $K_{eq} = \frac{k_1}{k_{-1}} = \frac{[Cl]^2}{[Cl_2]}$

Solving for [Cl] we get $[Cl] = K_{eq}^{1/2}\,[Cl_2]^{1/2}$

(2) Rate = $k_2\,[CHCl_3][Cl]$

Combining equations (1) and (2) by eliminating the active intermediate [Cl] and getting the rate law in terms of the reactants we get

Rate = $k_2\,K_{eq}^{1/2}\,[CHCl_3][Cl_2]^{1/2}$

11.43 According to the mechanism above, the rate law will be

a. $k_2 K_{eq} \frac{[O_3]^2}{[O_2]}$ b. $K_{eq} [O_3]$ c. $k_2[O][O_3]$ d. $K_{eq} \frac{[O_2][O]}{[O_3]}$

e. $K_{eq} k_2 [O_3]^2 [O]$

11.44 When the concentration of O_2 is doubled, the rate will

a. increase b. decrease c. will not change d. will double

e. will quadruple

11.45 The rate law for the reaction $2NO(g) + O_2(g) \rightarrow 2NO_2(g)$ is found to be:

$$RATE = k[NO]^2[O_2]$$

All the following mechanism are consistent with this rate law EXCEPT

a. $NO(g) + O_2(g) \overset{K_{eq}}{\rightleftarrows} NO_3(g)$ fast

$NO_3(g) + NO(g) \rightarrow 2NO_2(g)$ slow

b. $NO(g) + NO(g) \overset{K_{eq}}{\rightleftarrows} N_2O_2(g)$ fast

$N_2O_2(g) + O_2(g) \rightarrow 2NO_2(g)$ slow

c. $2NO(g) + O_2(g) \rightarrow 2NO_2(g)$ slow

d. $NO(g) + O_2(g) \rightarrow NO_2(g) + O(g)$ slow

$NO(g) + O(g) \rightarrow NO_2(g)$ fast

e. $2NO(g) + O_2(g) \rightarrow N_2O_4(g)$ slow

$N_2O_4(g) \rightarrow 2NO_2(g)$ fast

11.43 (a) Use the rapid equilibrium step and solve for the [O] concentration.

$$K_{eq} = \frac{[O_2][O]}{[O_3]} \quad ; \quad [O] = \frac{K_{eq}[O_3]}{[O_2]}$$

The rate of the reaction depends upon the rate of the slow step. Rate = k $[O][O_3]$

We would like to eliminate any intermediates from the experimental rate law and try to obtain a rate law in terms of the reactants. Combining the two equations above by eliminating [O], we get

$$\text{Rate} = \frac{k_2K_{eq}[O_3]^2}{[O_2]}$$

11.44 (b) One solves this question by using the experimental rate law from question 11.43. Since O_2 is in the denominator, increasing the O_2 concentration would decrease the rate of the reaction.

11.45 (d) Mechanism (d) is not consistent with the rate law because the rate law for that mechanism is Rate = $k[NO][O_2]$. This is a second order rate law not a third order. I will show that mechanism (a) gives the proper rate law.

(1) $K_{eq} = \frac{[NO_3]}{[NO][O_2]} \quad ; \quad [NO_3] = K_{eq}[NO][O_2]$

(2) Rate = $k[NO_3][NO]$

Combining equations (1) and (2) by eliminating $[NO_3]$, we get Rate = $k_2K_{eq}\ [NO]^2[O_2]$. Using a similar procedure, mechanism (b) can be shown to give a third order rate low. The slow step in mechanisms (c) and (e) shows a third order dependency and there are no steps before the slow step that need to be taken into consideration.

EQUILIBRIUM

Gaseous Equilibria

12.1 If the equilibrium constant for $A(g) + 2B \rightleftarrows 3C(g)$ is 2.25, then the concentration of A in equilibrium with 2.00 moles per liter of B and 3.00 moles per liter of C is

a. 1.00 M b. 2.00 M c. 2.25 M d. 3.00 M e. 4.00 M

12.2 At 30°C an equilibrium mixture of $N_2O_4(g)$ and $NO_2(g)$ exerted a total pressure of 0.750 atm. The partial pressure of N_2O_4 was 0.500 atm. K_p for the reaction $N_2O_4(g) \rightleftarrows 2\ NO_2(g)$ is

a. 1.50 b. 1.00 c. 0.667 d. 0.500 e. 0.125

12.3 The reaction of nitrogen with hydrogen to produce ammonia is represented by the following equilibrium:

$$N_2(g) + 3H_2(g) \rightleftarrows 2NH_3(g)$$

at 750 K, $K_c = 0.010$. If the concentration of NH_3 is 10^{-3} M and N_2 is 0.10 M, then the concentration of H_2 is

a. 1.0 M b. 0.10 M c. 0.010 M d. 0.0010 M e. 0.00010 M

12.4 At 425°C, a 1.00 L vessel contained 0.100 moles of I_2, 0.100 moles of H_2, and 0.700 moles of HI in equilibrium. K_c for the reaction, $H_2(g) + I_2(g) \rightleftarrows 2HI(g)$, is

a. 70.0 b. 49.0 c. 14.0 d. 6.90 e. 2.70

12.5 Given the equilibrium, $2SO_2(g) + O_2(g) \rightleftarrows 2SO_3(g)$, if this equilibrium is established by adding equal number of moles of SO_2 and O_2 to a 1.0 L bulb, then the following must be true at equilibrium:

a. $[SO_2] = [SO_3]$ b. $2[SO_2] = 2[SO_3]$ c. $[SO_2] = [O_2]$
d. $[SO_2] < [O_2]$ e. $[SO_2] > [O_2]$

12.1 (d) $K = \frac{[C]^3}{[A][B]^2} = 2.25$

$$[A] = \frac{[3]^3}{2.25\ [2]^2} = 3.00$$

60% 12.2 (e) $P_{NO_2} = P_T - P_{N_2O_4} = 0.750 - 0.500 = 0.250$ atm

$$K_P = \frac{P^2_{NO_2}}{P_{N_2O_4}} = \frac{(0.250)^2}{0.500} = 0.125$$

45% 12.3 (b) $K = \frac{[NH_3]^2}{[N_2][H_2]^3}$

$$H_2 = \left(\frac{[10^{-3}]^2}{0.010\ [0.10]}\right)^{1/3} = 0.10\ M$$

12.4 (b) $K = \frac{[HI]^2}{[H_2][I_2]} = \frac{[0.7]^2}{[0.1][0.1]} = 49$

48% 12.5 (d) SO_2 is the limiting reagent (O_2 is in excess).

12.6 Dinitrogen tetroxide readily undergoes decomposition to form red-brown $NO_2(g)$ as represented by the equation $N_2O_4(g) \rightleftarrows 2\ NO_2(g)$. At 25°C, 0.11 mole of N_2O_4 reacts to form 0.10 mole of N_2O_4 and 0.02 mole of NO_2. At 90°C, 0.11 mole of N_2O_4 forms 0.050 mole of N_2O_4 and 0.12 mole of NO_2. From these data we can conclude

a. N_2O_4 molecules react by a second order rate law.
b. N_2O_4 molecules react by a first order rate law.
c. the reaction above is exothermic.
d. N_2O_4 molecules react faster at 25°C than at 90°C.
e. the equilibrium constant for the reaction above increases with an increase in temperature.

12.7 For the equilibrium system

$$H_2O(g) + CO(g) \rightleftarrows H_2(g) + CO_2(g) \qquad \Delta H = -42\ \frac{kJ}{mol}$$

K equals 0.62 at 1260 K. If 0.10 mol each of H_2O, CO, H_2 and CO_2 (all at 1260 K) were placed in a 1.0 L thermally insulated vessel which was also at 1260 K, then when the system came to equilibrium

a. the temperature would decrease and the mass of CO would increase.
b. the temperature would decrease and the mass of CO would decrease.
c. the temperature would remain constant and the mass of CO would increase.
d. the temperature would increase and the mass of CO would decrease.
e. the temperature would increase and the mass of CO would increase.

12.6 (e) $N_2O_4(g) + \text{heat} \rightleftarrows 2\ NO_2(g)$

The reaction is endothermic so an increase in temperature favors the formation of more product.

30% 12.7 (a) $Q \text{ (trial product)} = \dfrac{(H_2)(CO_2)}{(H_2O)(CO)}$

$$= \frac{(0.10)(0.10)}{(0.10)(0.10)} = 1$$

Since Q is greater than K, more reactants need to be formed from the products. Heat will be consumed in producing the reactants and the temperature would decrease.

The information below is for questions 12.8 and 12.9. At a given temperature, 0.300 mol NO, 0.200 mol Cl_2 and 0.500 mol ClNO were placed in a 25.0 L container. The following equilibrium is established:

$$2\ ClNO(g) \rightleftarrows 2\ NO(g) + Cl_2(g)$$

12.8 At equilibrium, 0.600 mol of ClNO was present. The number of moles of Cl_2 present at equilibrium is

a. 0.050 b. 0.100 c. 0.150 d. 0.200 e. 0.250

12.9 The value of the equilibrium constant K_c is

a. 4.45×10^{-4} b. 6.67×10^{-4} c. 0.111 d. 0.167 e. 1500

Questions 12.10 and 12.11 refer to the following equilibrium at a given temperature:

$$4\ NH_3(g) + 3\ O_2(g) \rightleftarrows 2\ N_2(g) + 6\ H_2O(g)$$

12.10 Suppose the equilibrium above were established by adding 0.30 mole of NH_3 and 0.40 mole of oxygen to a 5.0 ℓ container. If y moles of water are present at equilibrium, then the number of moles of ammonia remaining at equilibrium would be:

a. 0.30 - y b. 0.40 - 3/2 y c. 4/6 y d. 0.30 - 3/2 y
e. 0.30 - 2/3 y

12.11 Under the conditions above, the equilibrium concentration of oxygen will be

a. 0.40 - 0.50 y b. 0.30 - 0.50 y c. 0.08 - 0.10 y
d. 0.06 - 0.13 y e. 0.40 - 3/2 y

68% 12.8 (c) $2ClNO \rightarrow 2NO + Cl_2$

	ClNO	NO	Cl_2
initial	0.500 mol	0.300 mol	0.200 mol
equil.	0.600 mol	0.300-0.100	0.200-0.050

mol Cl_2 needed to produce 0.100 mol more of ClNO is

$$\# \text{ mol } Cl_2 = \frac{1\ Cl_2}{2ClNO} \times 0.100 \text{ mol ClNO} = 0.050 \text{ mol}$$

$\therefore$ # mol Cl_2 at equilibrium = 0.200-0.050 = 0.150

72% 12.9 (b) We need molar concentrations.

$\therefore [ClNO] = \dfrac{0.600}{25.0} = 0.0240$; $[Cl_2] = \dfrac{0.150}{25.0} = 0.00600$;

$[NO] = \dfrac{0.200}{25.0} = 0.00800$

$$K_c = \frac{[NO]^2[Cl_2]}{[ClNO]^2} = \frac{[0.00800]^2[0.00600]}{[0.0240]^2} = 6.67 \times 10^{-4}$$

12.10 (e) $4\ NH_3 + 3\ O_2 \rightleftarrows 2\ N_2 + 6\ H_2O$

	NH_3	O_2	N_2	H_2O
Initial Amts.:	0.30 mol	0.40 mol	0	0
Equil. Amts.:	0.30 - 2/3 y			y

$\#$ mol NH_3 to produce y $H_2O = \dfrac{4\ NH_3}{6\ H_2O} \times y\ H_2O = 2/3\ y\ NH_3$

mol NH_3 at equilibrium = 0.30 - 2/3 y

$\#$ mol O_2 need to react $= \dfrac{3\ O_2}{6\ H_2O} \times y\ H_2O = 0.5\ y\ O_2$

mol O_2 at equilibrium = 0.40 - 0.5 y

12.11 (c) $\dfrac{0.40 - 0.5\ y \text{ mol}}{5.0\text{ L}} = (0.08 - 0.10\ y)\ \dfrac{\text{mol}}{\text{L}}$

12.12 The process, $2NH_3(g) + \text{heat} \rightleftarrows 3H_2(g) + N_2(g)$, is at equilibrium at a given temperature. Which of the following changes will change the EQUILIBRIUM CONSTANT for the process?

a. Adding more NH_3 to the system (volume held constant).
b. Increasing the pressure by reducing the volume of the system.
c. Adding an effective catalyst for the process.
d. Raising the temperature of the system.
e. Adding some neon to the system.

12.13 Which of the following equilibria would not be affected by volume changes at constant temperature?

a. $2CO(g) + O_2(g) \rightleftarrows 2CO_2(g)$
b. $2NO_2(g) \rightleftarrows N_2O_4(g)$
c. $2NO(g) + 3F_2(g) \rightleftarrows 2F_3NO(g)$
d. $BF_3(g) + NH_3(g) \rightleftarrows H_3NBF_3(s)$
e. $O_3(g) + NO(g) \rightleftarrows NO_2(g) + O_2(g)$

12.14 For which of the following systems at equilibrium and at constant temperature will decreasing the volume cause the equilibrium to shift to the right?

a. $2NO(g) + O_2(g) \rightleftarrows 2NO_2(g)$
b. $2H_2O(g) \rightleftarrows 2\ H_2(g) + O_2(g)$
c. $2SO_3(g) \rightleftarrows 2SO_2(g) + O_2(g)$
d. $CaCO_3(s) \rightleftarrows CaO(s) + CO_2(g)$
e. $H_2(g) + F_2(g) \rightleftarrows 2HF(g)$

12.15 In which of the following reactions does an instantaneous increase in the volume of the reaction vessel favor formation of the products?

a. $CO(g) + Cl_2(g) \rightleftarrows COCl_2(g)$
b. $PF_3Cl_2(g) \rightleftarrows PF_3(g) + Cl_2(g)$
c. $H_2(g) + I_2(g) \rightleftarrows 2HI(g)$
d. $N_2(g) + O_2(g) \rightleftarrows 2NO(g)$
e. $N_2(g) + 3H_2(g) \rightleftarrows 2NH_3(g)$

12.12 (d) A catalyst and addition of neon will not shift an equilibrium constant. Adding or removing a reactant or product will shift the equilibrium but will have no effect upon the equilibrium constant.

72% 12.13 (e) Δn_{gas} = # mol gaseous products - # mol gaseous reactants. When Δn_{gas} = zero, a volume change doesn't affect the concentrations of the products any differently than the reactants so there is no shift in equilibrium.

12.14 (a) See the solution to question 12.13.

$\Delta n_{gas} = 2 - 3 = -1$

When $\Delta n_{gas} \neq 0$, a volume change will cause a shift in the equilibrium. The shift will occur in such a way as to minimize the concentration effect (the equilibrium shifting to the side that has the fewest molecules).

12.15 (b) See questions 12.13 and 12.14. The equilibrium will shift to the side that has the most molecules.

$\Delta n_{gas} = 2 - 1 = +1$

12.16 For the reaction system, $N_2(g) + 3H_2(g) \rightleftarrows 2NH_3(g)$ + heat, the conditions which would favor maximum conversion of the reactants to products would be

a. high temperature and high pressure.

b. high temperature, pressure unimportant.

c. high temperature and low pressure.

d. low temperature and high pressure.

e. low temperature and low pressure.

12.17 In a chemical reaction at constant temperature, the addition of a catalyst

a. affects the heat content of the reaction.

b. affects the magnitude of the forward and reverse rate constants differently.

c. increases the energy of the molecules.

d. decreases the energy released in the chemical reaction.

e. increases the energy released in the chemical reaction.

12.18 Given the equilibrium, $PCl_3(g) + Cl_2(g) \rightleftarrows PCl_5(g)$ $\Delta H = -92$ kJ, the concentration of Cl_2 at equilibrium will be increased by

a. increasing the pressure.

b. adding PCl_3 to the system.

c. decreasing the temperature.

d. addition of helium.

e. addition of PCl_5.

12.19 Solid HgO, liquid Hg, and gaseous O_2 are placed in a glass bulb and are allowed to reach equilibrium at a given temperature.

$$2HgO(s) \rightleftarrows 2Hg(\ell) + O_2(g) \quad \Delta H = 43.4 \text{ kcal}$$

The mass of HgO in the bulb could be increased by

a. adding more Hg. b. removing some O_2.

c. reducing the volume of the bulb. d. increasing the temperature.

e. removing some Hg.

12.16 (d) Since the reaction is exothermic, lowering the temperature will shift the equilibrium to the right. Increasing the concentrations will shift the equilibrium in the direction of the fewest molecules.

12.17 (b) A catalyst only effects the rate of a reaction as the barrier (the activation energy) is changed. A different pathway is provided for the molecules in both the forward reaction and the reverse reaction.

12.18 (e) Responses (a), (b) and (c) would all cause the equilibrium to shift to the right. Addition of helium would have no effect.

12.19 (c) For this equilibrium the equilibrium expression is $K = [O_2]$ Reducing the volume of the bulb would increase the concentration of the oxygen.

12.20 If the system, $H_3NBF_3(s) \rightleftarrows NH_3(g) + BF_3(g)$, is at equilibrium at constant temperature, and the number of moles of H_3NHF_3 in the vessel is doubled then

a. the equilibrium amount of NH_3 and BF_3 is increased.

b. the amount of NH_3 and BF_3 is doubled.

c. the partial pressure of both NH_3 and BF_3 increases until equilibrium is reached.

d. the amount of H_3NBF_3 will then start decreasing.

e. the partial pressure of NH_3 and BF_3 will remain the same.

Equilibrium Expressions and Constants

12.21 For the reaction system, $CoO(s) + H_2(g) = Co(s) + H_2O(g)$, at 550°C, K = 67. The equilibrium constant expression is

a. $\frac{[CoO][H_2]}{[Co][H_2O]}$ b. $\frac{[Co][H_2O]}{[CoO][H_2]}$ c. $\frac{[Co][H_2O]}{[H_2]}$ d. $\frac{[H_2]}{[H_2O]}$

e. $\frac{[H_2O]}{[H_2]}$

12.22 A mixture of NO and O_2 at a fixed temperature reacts according to the equation, $2NO(g) + O_2(g) \rightleftarrows 2NO_2(g)$. When equilibrium is reached, which of the following ratios would be constant regardless of the initial concentrations of NO and O_2?

a. $\frac{[NO_2]}{[NO]^2[O_2]}$ b. $\frac{[NO_2]^2}{2[NO][O_2]}$ c. $\frac{[NO_2]^2}{[NO]^2[O_2]}$ d. $\frac{2[NO_2]}{2[NO]+[O_2]}$

e. $\frac{[NO_2]^2}{[NO]^2+[O_2]}$

12.20 (e) For this system the equilibrium expression is

$$K = [NH_3][BF_3]$$

Addition of more solid will have no effect as $H_3NBF_3(s)$ does not appear in the equilibrium expression.

12.21 (e) We do not include the concentrations of pure solids in the equilibrium expression because their concentrations are constants.

90% 12.22 (c) The equilibrium expression is a constant when the concentrations of the reaction products, each raised to a power equal to its stoichiometric coefficient in the chemical reaction, appear in the numerator, and the concentrations of the reactants, each raised to the appropriate power, appear in the denominator.

12.23 For the reaction system, $H_2(g) + I_2(g) \rightleftarrows 2\ HI(g)$, K = 55 at 425°C. For the system, $HI(g) \rightleftarrows 1/2\ H_2(g) + 1/2\ I_2(g)$, at 425°C, K is

a. -55 b. 55 c. 0.13 d. 0.018 e. 3.3×10^{-4}

12.24 At 5000°K, K = 1 for the reaction, $1/2\ N_2 + 1/2O_2 \rightleftarrows NO$. At 5000°K, one can be sure that in all N_2, O_2 and NO systems that

a. $[N_2] = [O_2] = 1$ b. $[N_2] + [O_2] = 1$

c. $[1/2\ N_2] + [1/2\ O_2] = [NO]$ d. $[N_2] = [O_2] = [NO]$

e. $\dfrac{[NO]}{[N_2]^{1/2}[O_2]^{1/2}} = 1$

12.25 For which of the following reactions are the numerical values of K_p and K_c the same?

a. $2CO(g) + O_2(g) \rightleftarrows 2CO_2(g)$
b. $PCl_5(g) \rightleftarrows PCl_3(g) + Cl_2(g)$
c. $N_2(g) + 3H_2(g) \rightleftarrows 2NH_3(g)$
d. $2H_2O(g) \rightleftarrows 2H_2(g) + O_2(g)$
e. $H_2(g) + I_2(g) \rightleftarrows 2HI(g)$

12.26 For the reaction system, $N_2O_4(g) \rightleftarrows 2NO_2(g)$, at 90°C, $K_c = 0.27$. For this system K_p is

a. > 0.27 b. < 0.27 c. = 0.27 d. cannot determine without additional data.

12.23 (c) Two rules must be applied to answer this question. The equilibrium constant for a reaction and its reverse are reciprocals of each. Since this reaction is a multiple of the original, the reaction must be multiplied by one-half and its equilibrium constant must be raised to the one-half power.

$$K_2 = \left(\frac{1}{K_1}\right)^{1/2} = \left(\frac{1}{55}\right)^{1/2} = 0.13$$

12.24 (e) At a given temperature, the equilibrium expression must be equal to a constant.

12.25 (e) $K_p = K_c\ (RT)^{\Delta n}$ where Δn = the number of moles of gaseous products minus the number of moles of gaseous reactants in the balanced equation. Since $\Delta n = 0$ for equation (e), K_p will equal K_c.

12.26 (a) $K_p = K_c\ (RT)^{\Delta n}$ and $\Delta n = 1$

$$K_p = 0.27\ \frac{\text{mol}}{\text{L}}\ [(0.0821\ \text{L·atm/mol·K})(363\ \text{K})]^1$$

$$K_p = 8.0\ \text{atm}$$

12.27 Given the equilibrium constants for the following equilibria at 25°C

	K
$H_2S(aq) \rightleftarrows H^+(aq) + HS^-(aq)$	1.0×10^{-7}
$H^+(aq) + S^{2-}(aq) \rightleftarrows HS^-(aq)$	8.3×10^{12}

The equilibrium constant for the reaction, $H_2S(aq) \rightleftarrows 2H^+(aq) + S^{2-}(aq)$ is

a. 8.3×10^5 b. 9.3×10^5 c. 1.2×10^6 d. 2.2×10^{-20}

e. 1.2×10^{-20}

12.28 If K = 0.145 for $A_2 + 2B \rightleftarrows 2\ AB$ then for $AB \rightleftarrows B + 1/2\ A_2$, K would equal

a. 0.145 b. -0.145 c. 0.381 d. 2.63 e. 6.90

pH and pOH

12.29 The hydronium ion concentration of a 0.00100 acetic acid solution is 1.34×10^{-4} M. The pH of the solution is

a. 3.00 b. 3.40 c. 3.87 d. 4.00 e. 4.13

12.30 A solution has a pH value of 10.82. The $[OH^-]$ concentration for this solution is

a. 1.5×10^{-11} b. 1.5×10^{-10} c. 6.6×10^{-10}

d. 1.5×10^{-5} e. 6.6×10^{-4}

12.31 A solution has a hydroxide concentration equal to 4.8×10^{-3}. The pH for this solution is

a. 11.68 b. 8.92 c. 4.80 d. 3.68 e. 2.32

12.27 (e) $H_2S(aq) \rightleftarrows H^+(aq) + HS^-(aq)$ 1.0×10^{-7}

$\underline{HS^-(aq) \rightleftarrows H^+(aq) + S^{2-}(aq)}$ $\underline{1/8.3 \times 10^{12}}$

$H_2S)aq) \rightleftarrows 2H^+(aq) + S^{2-}(aq)$ 1.2×10^{-20}

52% 12.28 (d) See the solution to question 12.23.

$K = (1/0.145)^{1/2} = 2.63$

12.29 (c) $pH = -\log [H^+] = -\log [1.34 \times 10^{-4}] = 3.87$

92% 12.30 (e) $pOH = 14.00 - 10.82 = 3.18$

$[OH^-] = 10^{-pOH} = 10^{-3.18} = 6.6 \times 10^{-4}$

92% 12.31 (a) $pOH = -\log [OH^-] = -\log [4.8 \times 10^{-3}] = 2.32$
$pH = 14.00 - 2.32 = 11.68$

12.32 The pOH of a solution is 6.50. Its hydroxide ion concentration is

a. 5.0×10^{-6} b. 3.2×10^{-7} c. 2.5×10^{-6} d. 3.0×10^{-6}

e. None of these.

12.33 A solution which has the pH of 8.7 would be described as

a. very acidic b. slightly acidic c. neutral d. very basic

e. slightly basic

12.34 A solution which is formed by combining 150. ml of 0.20 M H_2SO_4 with 100. ml of 0.70 M NaOH has a $[H_3O]^+$ concentration of _______ and a pH of _______.

a. 0.040 and 1.4 b. 0.016 and 1.8 c. 1.0×10^{-12} M and 12.0

d. 2.5×10^{-13} M and 12.6 e. 6.25×10^{-14} M and 13.2

12.35 A solution which is formed by combining 300. ml of 0.10 M $Ba(OH)_2$ with 200. ml of 0.15 M HCl has an OH^- concentration of _______ and a pOH of _______.

a. 1×10^{-7} and 7.0 b. 0.03 M and 1.5 c. 0.06 M and 1.2

d. 3.3×10^{-13} M and 12.5 e. 6.7×10^{-13} M and 12.2

12.36 The pH of a 0.01 M HCOOH (formic acid) is approximately

a. 1 b. 3 c. 7 d. 11 e. 13

12.37 The pH of a 0.05 M solution of ammonia is approximately

a. 1 b. 3 c. 7 d. 11 e. 13

12.38 The ionization constant of water at 45°C is 4.0×10^{-14}. What is the pH of water (a neutral solution) at this temperature?

a. 5.0 b. 6.7 c. 7.0 d. 7.3 e. 13.4

12.32 (b) $[OH^-] = 10^{-pOH} = 10^{-6.50} = 3.2 \times 10^{-7}$

95% 12.33 (e) At 25°C a pH of 7.00 is a neutral solution. pH values greater than 7 correspond to basic solutions. Values near 7 would be slightly basic solutions.

12.34 (d) # mmol H^+ = 150 mL x 0.20 M x 2 H^+/1 H_2SO_4 = 60

mmol OH^- = 100 mL x 0.70 M = 70

mmol OH^- after reaction = 70-60 = 10

$$M_{OH^-} = \frac{10 \text{ mmol}}{250 \text{ mL}} = 0.040 \text{ M}$$

$$[H^+] = \frac{1.0 \times 10^{-14}}{0.040}$$

12.35 (c) $Ba(OH)_2 + 2\ HCl \rightarrow BaCl_2 + 2\ H_2O$

$$\#\text{ mmol } OH^- = 0.10 \text{ M} \times 300 \text{ mL} \times \frac{2\ OH^-}{1Ba(OH)_2} = 60$$

mmol H^+ = 0.15 M x 200 mL = 30

excess OH^- = 60-30 = 30 mmol

$$M_{OH^-} = \frac{30 \text{ mmol}}{300 \text{ mL} + 200 \text{ mL}} = 0.060 \text{ M}$$

12.36 (b) Formic acid is a weak acid. A strong acid is required in order to have pH values around one.

12.37 (d) Ammonia is a weak base. A strong base is required in order to have pH values around 13.

12.38 (b) $K_w = [H^+][OH^-] = 4.0 \times 10^{-14}$

$[H^+]^2 = 4.0 \times 10^{-14}$

$[H^+] = 2.0 \times 10^{-7}$; pH = $-\log [2.0 \times 10^{-7}] = 6.7$

Weak Acid and Weak Base Equilibria

12.39 The K_a (HCO_3^-) is the equilibrium constant for the reaction

a. $H_2CO_3 + H_2O \rightleftarrows H_3O^+ + HCO_3^-$

b. $HCO_3^- + H_2O \rightleftarrows H_3O^+ + CO_3^{2-}$

c. $HCO_3^- + H_2O \rightleftarrows H_2CO_3 + OH^-$

d. $HCO_3^- + H_3O^+ \rightleftarrows H_2CO_3 + H_2O$

e. $HCO_3^- + OH^- \rightleftarrows CO_3^{2-} + H_2O$

12.40 A 0.010 M solution of a weak monoprotic acid is 5.0% dissociated. The equilibrium constant, K_a, for this acid is

a. 5.0×10^{-2} b. 5.0×10^{-3} c. 2.5×10^{-4} d. 2.5×10^{-5}
e. 2.5×10^{-6}

12.41 A 0.10 M solution of weak monoprotic acid has a hydronium ion concentration of 5.0×10^{-4} M. The equilibrium constant, K_a, for this acid is

a. 5.0×10^{-2} b. 5.0×10^{-3} c. 5.0×10^{-4} d. 2.5×10^{-5}
e. 2.5×10^{-6}

12.42 Hydrofluoric acid has a dissociation constant at 25°C of 6.7×10^{-4}. If 1.0 mole of HF is dissolved in enough water to give 1.00 L of solution, what is the hydrogen ion concentration in this solution expressed in moles/liter?

a. 2.0×10^{-2} to 2.4×10^{-2} b. 2.5×10^{-2} to 2.7×10^{-2}
c. 3.0×10^{-3} to 3.5×10^{-3} d. 6.6×10^{-4} to 6.8×10^{-4}
e. None of the above.

72% 12.39 (b) K_a is when the solute reacts as an acid (a proton donor) with water.

12.40 (d) $HA + H_2O \rightleftarrows H_3O^+ + A^-$

$$[H_3O^+] = [A^-] = 0.010 \times 0.050 = 5.0 \times 10^{-4} \text{ M}$$

$$K_a = \frac{[H_3O^+][A^-]}{[HA]} = \frac{[5.0 \times 10^{-4}]^2}{[0.010-5.0 \times 10^{-4}} = 2.5 \times 10^{-5}$$

12.41 (e) $K_a = \frac{[5.0 \times 10^{-4}]^2}{[0.10]} = 2.5 \times 10^{-6}$

80% 12.42 (b) $HF + H_2O \rightleftarrows H_3O^+ + F^-$

$1.0 - y \qquad y \qquad y$

$$K_a = \frac{[H_3O^+][F^-]}{[HF]} = \frac{y^2}{1.0} = 6.7 \times 10^{-4}$$

$$y^2 = 6.7 \times 10^{-4}$$

$$y = [H_3O^+] = 2.6 \times 10^{-2}$$

12.43 Calculate the dissociation constant of an acid if a 1.00 M solution of the acid has a pH of 3.18.

a. 4.4×10^{-7} b. 4.4×10^{-6} c. 6.6×10^{-4} d. 6.6×10^{-3}
e. 3.2×10^{-3}

12.44 What is $[H_3O^+]$ for a 0.010 M HOBr solution? $K_a(HOBr) = 2.5 \times 10^{-9}$.

a. 2.5×10^{-11} M b. 2.5×10^{-7} M c. 5.0×10^{-7} M
d. 5.0×10^{-6} M e. 5.0×10^{-5} M

12.45 A 2.4×10^{-2} M solution of benzoic acid is 5.0% dissociated. In a 2.4×10^{-3} M solution, the percentage of dissociation would be

a. the same. b. > 5% c. < 5% d. zero e. none of these.

12.46 The symbol, $K_b\ (H_2PO_4^-)$, is the equilibrium constant for the reaction

a. $H_2PO_4^- + OH^- \rightleftarrows HPO_4^{2-} + H_2O$
b. $H_2PO_4^- + H_2O \rightleftarrows HPO_4^{2-} + H_3O^+$
c. $H_2PO_4^- + H_2O \rightleftarrows H_3PO_4 + OH^-$
d. $H_2PO_4^- + H_3O^+ \rightleftarrows H_3PO_4 + H_2O$
e. $H_2PO_4^- + 2H_2O \rightleftarrows 2H_3O^+ + PO_4^{3-}$

12.47 For which of the following reactions is the equilibrium constant called a basicity constant, K_b?

a. $HCOOH(aq) + NH_3(aq) \rightleftarrows HCOO^-(aq) + NH_4^+(aq)$
b. $H_3O^+(aq) + OH^-(aq) \rightleftarrows 2\ H_2O$
c. $HOCl(aq) + H_2O \rightleftarrows H_3O^+(aq) + OCl^-(aq)$
d. $CH_3COOH(aq) + OH^-(aq) \rightleftarrows CH_3COO^-(aq) + H_2O$
e. $CH_3NH_2(aq) + H_2O \rightleftarrows CH_3NH_3^+(aq) + OH^-(aq)$

60% 12.43 (a) $[H_3O^+] = 10^{-pH} = 10^{-3.18} = 6.6 \times 10^{-4}$

$$K_a = \frac{[H_3O^+][A^-]}{[HA]} = \frac{[6.6 \times 10^{-4}]^2}{[1.00]} = 4.4 \times 10^{-7}$$

12.44 (d) $HOBr + H_2O \rightleftarrows H_3O^+ + Br^-$

$0.010 - y \qquad y \qquad y$

$$\frac{y^2}{0.010 - \cancel{y}} = 2.5 \times 10^{-9}$$

$$y^2 = 2.5 \times 10^{-11}$$

$$y = [H_3O^+] = 5.0 \times 10^{-6}\ M$$

12.45 (b) The percentage dissociation increases as the concentration of the acid decreases.

70% 12.46 (c) $H_2PO_4^-$ (the solute) is an acid and transfers a proton to water.

84% 12.47 (e) K_b is when the solute is acting as a base (a proton acceptor) with water.

12.48 K_b for NH_3 is 2×10^{-5}. K_a for the NH_4^+ ion is

a. 2×10^{-5} b. 5×10^{-9} c. 5×10^{-10} d. 2×10^{-19}

12.49 A 1.00 L solution was prepared by combining 0.200 mol HCl with 0.100 mol CH_3COOH ($K_a = 1.8 \times 10^{-5}$). What is the equilibrium concentration of CH_3COO^-?

a. 3.6×10^{-6} M b. 9.0×10^{-6} M c. 3.6×10^{-5} M
d. 1.8×10^{-5} M e. 1.3×10^{-3} M

12.50 What is the pH of 0.200 M HCN ($K_a = 4.0 \times 10^{-10}$)?

a. 4.70 b. 5.05 c. 5.20 d. 5.70 e. 10.10

The information below is for the next two questions.

12.51 Calculate the hydronium ion concentration of a 0.0100 M solution of chloroacetic acid, $ClCH_2COOH$, ($K_a = 1.4 \times 10^{-3}$). Since the acid has a relatively large equilibrium constant and the solution is relatively dilute, it will be necessary to use either the quadratic equation or the method of successive approximation,

a. 0.0031 b. 0.0037 c. 0.0063 d. 0.0069 e. 0.0073

12.52 What is the equilibrium concentration of $ClCH_2COOH$ in a solution which is called 0.0100 M of $ClCH_2COOH$?

a. 0.0027 b. 0.0031 c. 0.0063 d. 0.0069 e. 0.0073

12.48 (c) $NH_4^+ + H_2O \rightleftarrows NH_3 + H_3O^+$

$$K_a = \frac{[NH_3][H_3O^+]}{[NH_4^+]}$$

Multiplying by one ($[OH^-]/[OH^-]$) we get

$$K_a = \frac{[NH_3][H_3O^+][OH^-]}{[NH_4^+][OH^-]} = \frac{K_w}{K_b} = \frac{1.0 \times 10^{-14}}{2 \times 10^{-5}} = 5 \times 10^{-10}$$

12.49 (b) $CH_3COOH + H_2O \rightleftarrows H_3O^+ + CH_3COO^-$

0.100-y   0.200+y   y

$$K_a = \frac{[0.200+\cancel{y}][y]}{[0.100-\cancel{y}]} = 1.8 \times 10^{-5} \quad ; \quad y = 9.0 \times 10^{-6}$$

The amount of CH_3COOH that has dissociated and the additional amount of H_3O^+ that has come from CH_3COOH can be disregarded since y is so small.

12.50 (b) $HCN + H_2O \rightleftarrows H_3O^+ + CN^-$

0.200-y   y   y

$$\frac{y^2}{0.200} = 4.0 \times 10^{-10} \; ; \; y = 8.9 \times 10^{-6}$$

$$pH = -\log [8.9 \times 10^{-6}] = 5.05$$

12.51 (a) $ClCH_2COOH + H_2O \rightleftarrows H_3O^+ + ClCH_2COO^-$

0.0100-y   y   y

$$K_a = \frac{[y][y]}{[0.0100-y]} = 1.4 \times 10^{-3}$$

$$y^2 + 1.4 \times 10^{-3}\, y - 1.4 \times 10^{-5} = 0$$

$$y = \frac{-1.4 \times 10^{-3} \pm \sqrt{(1.4 \times 10^{-3})^2 - 4\,(1)\,-1.4 \times 10^{-5})}}{2}$$

$$y = 3.1 \times 10^{-3}$$

12.52 (d) $[ClCH_2COOH] = C_o - [H_3O^+] = 0.0100 - 3.1 \times 10^{-3} = 0.0069$ M

Polyprotic Acids

The information below may be useful for questions 12.53-12.58. Phosphoric acid is a triprotic acid, H_3PO_4, with $K_1 = 7.5 \times 10^{-3}$, $K_2 = 6.2 \times 10^{-8}$, and $K_3 = 3.6 \times 10^{-13}$.

12.53 What is the $[H_3O^+]$ concentration in a 0.25 M solution of H_3PO_4? This solution has a pH value of 1.40.

a. 0.027 b. 0.040 c. 0.043 d. 0.053 e. 0.0075

12.54 What is the $[H_2PO_4^-]$ concentration?

a. 0.040 b. 0.043 c. 0.053 d. 0.087 e. 0.0075

12.55 What is the % dissociation of the H_3PO_4?

a. 20 b. 19 c. 17 d. 16 e. 12

12.56 What is the $[HPO_4^{2-}]$ concentration?

a. 3.6×10^{-13} b. 6.2×10^{-8} c. 7.5×10^{-3} d. 1.55×10^{-8}
e. 1.24×10^{-4}

12.57 What is the $[PO_4^{3-}]$ concentration?

a. 1.5×10^{-10} b. 3.6×10^{-13} c. 5.6×10^{-19} d. 3.6×10^{-21}
e. 4.2×10^{-22}

12.58 What is the equilibrium constant for the reaction

$$H_3PO_4 + 3\ H_2O \rightleftarrows 3\ H_3O^+ + PO_4^{3-}\ ?$$

a. 3.6×10^{-13} b. 2.2×10^{-20} c. 2.8×10^{12} d. 1.7×10^{-22}
e. 6.0×10^{21}

12.53 (b) $[H_3O^+] = 10^{-pH} = 10^{-1.40}$

12.54 (a) $[H_3O^+] = [H_2PO_4^-]$

12.55 (d) $\frac{0.040}{0.25} \times 100$

12.56 (b) $H_2PO_4^- + H_2O \rightleftarrows HPO_4^{2-} + H_3O^+$

$$K_2 = \frac{[H_3O^+][HPO_4^{2-}]}{[H_2PO_4^-]} \quad ; \text{ since } [H_3O^+] = [H_2PO_4^-]$$

$$K_2 = [HPO_4^{2-}]$$

12.57 (c) $K_3 = \frac{[H_3O^+][PO_4^{3-}]}{[HPO_4^{2-}]}$

$$3.6 \times 10^{-13} = \frac{[0.040][PO_4^{3-}]}{[6.2 \times 10^{-8}]}$$

12.58 (d) $K_1K_2K_3$

12.59 H_3PO_4 is a ______ acid and H_3PO_4 is a ______ acid than $H_2PO_4^-$ and $H_2PO_4^-$ is a ______ acid than HPO_4^{2-}.

a. strong, stronger, stronger b. weak, weaker, weaker

c. weak, stronger, stronger d. strong, weaker, weaker

e. strong, weaker, stronger

12.60 In a 0.01 M solution of 1,4 butanedicarboxylic acid, $HOOCCH_2CH_2COOH$ ($K_1 = 2.9 \times 10^{-5}$, $K_2 = 5.3 \times 10^{-6}$), the species present in the lowest concentration is

a. $H_3O^+(aq)$ b. $HOOCCH_2CH_2COOH(aq)$ c. $HOOCCH_2CH_2COO^-(aq)$

d. $^-OOCCH_2CH_2COO^-(aq)$ e. H_2O

12.61 In a 0.01 M solution of 1,4 butanedicarboxylic acid, $HOOCCH_2CH_2COOH$ ($K_1 = 2.9 \times 10^{-5}$, $K_2 = 5.3 \times 10^{-6}$), the species present in the highest concentration is

a. $H_3O^+(aq)$ b. $HOOCCH_2CH_2COOH(aq)$ c. $HOOCCH_2CH_2COO^-(aq)$

d. $^-OOCCH_2CH_2COO^-(aq)$ e. $OH^-(aq)$

The information below may be useful for questions 12.62 - 12.64. Sulfurous acid is a diprotic acid, H_2SO_3, with $K_1 = 1.3 \times 10^{-2}$ and $K_2 = 5.6 \times 10^{-8}$. $K_w = 1.0 \times 10^{-14}$. In a 0.10 molar solution the pH is 1.52.

12.62 What is the $[H_3O^+]$ concentration in a sulfurous acid solution which is 0.10 M?

a. 0.036 b. 0.030 c. 1.3×10^{-2} d. 0.10 e. 0.070

12.63 What is the $[H_2SO_3]$ concentration in a sulfurous acid solution which is 0.10 M?

a. 0.10 b. 0.050 c. 0.030 d. 0.070 e. 0.064

50% 12.59 (c) Relative acidities for polyprotic acids are as follows: $H_3X > H_2X^- > HX^{2-}$. Phosphoric acid is a relatively weak acid.

12.60 (d) For a weak polyprotic acid, the negative ion produced by the last dissociation step will have the lowest concentration. See question 12.57.

80% 12.61 (b) Weak acids like 1,4 butanedicarboxylic acid are dissociated about 1% (of course the amount of disociation depends upon K as well as the concentration).

68% 12.62 (b) We need to get the $[H_3O^+]$ concentration from the pH otherwise we would have to use the quadratic equation or successive approximations.

$$[H_3O^+] = 10^{-pH} = 10^{-1.52} = 0.030$$

36% 12.63 (d) $[H_2SO_3] = C_o - [H_3O^+]$

$$= 0.100 - 0.030 = 0.070$$

12.64 What is the equilibrium constant for the reaction

$$H_2SO_3 + 2H_2O \rightleftharpoons 2H_3O^+ + SO_3^{2-} \quad ?$$

a. 1.3×10^{-2} b. 5.6×10^{-8} c. 1.8×10^{-7} d. 5.6×10^{-9}
e. 7.3×10^{-10}

The information below may be useful for questions 12.65 - 12.69
Arsenic acid is a triprotic acid, H_3AsO_4, with $K_1 = 2.5 \times 10^{-4}$, $K_2 = 5.6 \times 10^{-8}$, and $K_3 = 3.0 \times 10^{-13}$.

12.65 What is the $[H_3O^+]$ concentration in a 0.10 M solution if the percent dissociation is 5.0%?

a. 2.5×10^{-4} M b. 2.5×10^{-3} M c. 5.0×10^{-3} M
d. 5.0×10^{-2} M e. 5.0×10^{-1} M

12.66 What is the $[H_2AsO_4^-]$ concentration?

a. 2.5×10^{-4} M b. 2.5×10^{-3} M c. 5.0×10^{-3} M
d. 5.0×10^{-2} M e. 5.0×10^{-1} M

12.67 What is the $[HAsO_4^{2-}]$ concentration?

a. 2.5×10^{-4} M b. 5.6×10^{-8} M c. 5.6×10^{-9} M
d. 1.4×10^{-11} M e. 3.0×10^{-13} M

12.68 What is the $[AsO_4^{3-}]$ concentration?

a. 1.7×10^{-7} M b. 1.3×10^{-10} M c. 3.0×10^{-13} M
d. 2.5×10^{-16} M e. 3.4×10^{-18} M

12.69 What is the equilibrium constant for the reaction

$$H_3AsO_4 + 2H_2O \rightleftharpoons 2H_3O^+ + HAsO_4^{2-} \quad ?$$

a. 4.2×10^{-24} b. 1.7×10^{-20} c. 7.5×10^{-17}
d. 1.4×10^{-11} e. 5.6×10^{-8}

92% 12.64 (e) $H_2SO_3 + H_2O \rightleftarrows H_3O^+ + HSO_3^-$ $\quad K_1 = 1.3 \times 10^{-2}$

$HSO_3^- + H_2O \rightleftarrows H_3O^+ + SO_3^{2-}$ $\quad K_2 = 5.6 \times 10^{-8}$

$H_2SO_3 + 2H_2O \rightleftarrows 2H_3O^+ + SO_3^{2-}$ $\quad K_1K_2 = 7.3 \times 10^{-10}$

92% 12.65 (c) $[H_3O^+] = C_o \times$ Fraction of Dissociation

$= 0.10 \times 0.050 = 5.0 \times 10^{-3}$ M

92% 12.66 (c) $H_3AsO_4 + H_2O \rightleftarrows H_3O^+ + H_2AsO_4^-$

$0.10 - y \qquad y \qquad y$

$[H_3O^+] = [H_2AsO_4^-] = 5.0 \times 10^{-3}$ M

82% 12.67 (b) $H_2AsO_4^- + H_2O \rightleftarrows H_3O^+ + HAsO_4^{2-}$

$5.0 \times 10^{-3} - w \qquad 5.0 \times 10^{-3} \qquad w$

$$K_2 = \frac{[H_3O^+][HAsO_4^{2-}]}{[H_2AsO_4^-]}$$

$$5.6 \times 10^{-8} = \frac{[5.0 \times 10^{-3}][w]}{[5.0 \times 10^{-3} - w]} = 5.6 \times 10^{-8} = [HAsO_4^{2-}]$$

68% 12.68 (e) $HAsO_4^{2-} + H_2O \rightleftarrows H_3O^+ + AsO_4^{3-}$

$5.6 \times 10^{-4} - z \qquad 5.0 \times 10^{-3} \qquad z$

$$K_3 = \frac{[H_3O^+][AsO_4^{3-}]}{[HAsO_4^{2-}]} = 3.0 \times 10^{-13} = \frac{5.0 \times 10^{-3}][z]}{[5.6 \times 10^{-8} - z]}$$

$$z = [AsO_4^{3-}] = \frac{5.6 \times 10^{-8} \times 3.0 \times 10^{-13}}{5.0 \times 10^{-3}} = 3.4 \times 10^{-18} \text{ M}$$

72% 12.69 (d) $H_3AsO_4 + H_2O \rightleftarrows H_3O^+ + H_2AsO_4^-$ $\quad K_1 = 2.5 \times 10^{-4}$

$H_2AsO_4^- + H_2O \rightleftarrows H_3O^+ + HAsO_4^{2-}$ $\quad K_2 = 5.6 \times 10^{-8}$

$H_3AsO_4 + 2H_2O \rightleftarrows 2H_3O^+ + HAsO_4^{2-}$ $\quad K_1K_2 = 1.4 \times 10^{-11}$

Buffers, Titration Curves, Indicators and K_a Determination

12.70 Which one of the following mixtures will be a buffer when dissolved in a liter of water?

a. 0.1 mole of $Ba(OH)_2$ and 0.2 mole of HBr.

b. 0.3 mole of KCl and 0.3 mole of HCl.

c. 0.4 mole of NH_3 and 0.4 mole of HCl.

d. 0.2 mole of CH_3COOH and 0.1 mole of NaOH.

d. 0.2 mole of HBr and 0.1 mole of NaOH.

12.71 Which one of the following mixtures will be a buffer when dissolved in a liter of water?

a. 0.2 mole of NaCl and 0.1 mole of HCl.

b. 0.1 mole of NH_3 and 0.2 mole of HCl.

c. 0.1 mole of NaOH and 0.1 mole of HCl.

d. 0.2 mole of Na_2HPO_4 and 0.1 mole of HCl.

e. 0.2 mole of H_2SO_3 and 0.1 mole of HCl.

12.72 Which of the following mixtures will be a buffer solution when dissolved in 1.00 L of water?

a. 0.20 mol of NH_3 and 0.20 mol HCl.

b. 0.20 mol of NH_3 and 0.40 mol KOH.

c. 0.20 mol of KCl and 0.10 mol HCl.

d. 0.20 mol of KCl and 0.10 mol KOH.

e. 0.20 mol of NH_3 and 0.10 mol HCl.

12.73 An effective NaH_2PO_4 - Na_2HPO_4 buffer can be achieved by adding equal volumes of the following solutions EXCEPT

a. 1.0 M H_3PO_4 + 1.0 M Na_3PO_4

b. 1.0 M H_3PO_4 + 1.0 M NaH_2PO_4

c. 2.0 M NaH_2PO_4 + 1.0 M NaOH

d. 2.0 M Na_2HPO_4 + 1.0 M HCl

e. 1.0 M NaH_2PO_4 + 1.0 M Na_2HPO_4

12.70 (d) A buffer is a salt of a weak acid plus a weak acid or salt of a weak base plus a weak base. It is necessary to have both components. After the reaction of NaOH with CH_3COOH there are present aboth CH_3COOH and $NaCH_3COO$. Answer (c) could be a buffer system but there is only the salt and no weak base left over after reaction.

68% 12.71 (d) (a) Salt of a strong acid plus strong acid.

(b) NH_3 + HCl produces NH_4Cl + HCl (left over).

(c) NaCl is produced.

(d) After the reaction of HCl with Na_2HPO_4, there is present both NaH_2PO_4 and Na_2HPO_4.

(e) A weak acid and strong acid.

A buffer is a salt of a weak acid and a weak acid (with both present in reasonable concentrations).

32% 12.72 (e) (a) $NH_3 + HCl \rightarrow NH_4Cl$ (salt only)

(b) a weak base plus a strong base

(c) a salt plus a strong acid

(d) a salt plus a strong base

(e)

	NH_3 +	HCl →	NH_4Cl
initial	0.20 mol	0.10 mol	0 mol
final	0.10 mol	0 mol	0.10 mol

12.73 (b) (a)

	H_3PO_4 +	PO_4^{3-} →	$H_2PO_4^-$ +	$H_2PO_4^{2-}$
initial	1.0 mol	1.0 mol	0	0
final	0	0	1.0 mol	1.0 mol

After reaction, the H_3PO_4 - PO_4^{3-} system produces a $H_2PO_4^-$ - HPO_4^{2-} buffer system. Likewise, after reaction in systems (c) and (d), the desired buffer is produced.

(b) H_3PO_4 and $H_2PO_4^-$ is a buffer system but the wrong one.

12.74 If one were to prepare a buffer by using HSO_3^- and SO_3^{2-}, its pH would probably be in the range of (K_a HSO_3^- = 3×10^{-8})

a. 6.5 - 8.5 b. 10 - 12 c. 2.0 - 4.0 d. 4.5 - 6.5 (e) 8 - 10

12.75 The pH of a solution 0.050 M in HF and also 0.050 M in NaF is (K_a HF is 7.2×10^{-4})

a. 3.14 b. 4.13 c. 7.00 d. 9.87 e. 10.86

12.76 What is the hydronium ion concentration of a solution which is 0.010 M in formic acid, HCOOH, and 0.10 M in sodium formate. The K_a for HCOOH is 1.8×10^{-4}.

a. 1.8×10^{-7} M b. 1.8×10^{-6} M c. 1.8×10^{-5} M

b. 1.8×10^{-4} M e. 1.8×10^{-3} M

12.77 What is the hydrogen ion concentration of a solution which is 0.010 M with respect to acetic acid and 0.10 M in sodium acetate? The ionization constant of acetic acid is 1.8×10^{-5}.

a. 1.8×10^{-7} M b. 1.8×10^{-6} M c. 1.8×10^{-5} M

d. 1.8×10^{-4} M e. 1.8×10^{-3} M

12.78 What is the pH of a solution which is 0.0100 M in HA and also 0.0020 M in NaA (K_a = 9.0×10^{-6})?

a. 2.40 b. 4.35 c. 5.65 d. 6.65 e. 7.15

12.79 In titrating 0.20 M acetic acid, CH_3COOH, with 0.20 M NaOH, the solution at the equivalence point is

a. 0.20 M $NaCH_3COO$ b. neutral c. slightly acidic

d. 0.10 M CH_3COOH and 0.10 M NaOH e. 0.10 M $NaCH_3COO$

12.80 A buffer that is a mixture of ammonia (K_b = 1.8×10^{-5}) and ammonium sulfate has $[OH^-]$ = 6×10^{-6} M. The $[NH_4^+]$ / $[NH_3]$ ratio in this buffer is

a. 1:2 b. 2:1 c. 1:3 d. 3:1 e. 2:3

12.74 (a) $HSO_3^- + H_2O \rightleftarrows H_3O^+ + SO_3^{2-}$

$$K = \frac{[H_3O^+][SO_3^{2-}]}{[HSO_3^-]} \quad ; \quad [H_3O^+] = K\frac{[HSO_3^-]}{[SO_3^{2-}]}$$

For an effective buffer system, the ratio of acid to salt should be between 0.1 and 10. A buffer which is equally effective to the addition of both a strong acid and a strong base is one in which $[HSO_3^-] = [SO_3^{2-}]$. Then $[H_3O^+] = K$ and the pH is 7.5.

12.75 (a) This is a buffer in which the concentration of the salt is the same as the concentration of the acid so $K_a = [H_3O^+]$

$\therefore$ pH = -log [7.2 x 10^{-4}]

68% 12.76 (c) $HCOOH + H_2O \rightleftarrows H_3O^+ + HCOO^-$ $\qquad K_a = \frac{[H_3O^+][HCOO^-]}{[HCOOH]}$

$$[H_3O^+] = \frac{K_a[HCOOH]}{[HCOO^-]} = \frac{1.8 \times 10^{-4} \times 0.010}{0.10}$$

12.77 (b) $CH_3COOH + H_2O \rightleftarrows H_3O^+ + CH_3COO^-$

$$K_a = \frac{[H_3O^+][CH_3COO^-]}{[CH_3COOH]} = 1.8 \times 10^{-5}$$

$$[H_3O^+] = 1.8 \times 10^{-5} \times \frac{0.010}{0.10} = 1.8 \times 10^{-6}$$

12.78 (b) $\frac{[H_3O^+][A^-]}{[HA]} = 9.0 \times 10^{-6}$

$$[H_3O^+] = \frac{9.0 \times 10^{-6} \times 0.0100}{0.0020} = 4.5 \times 10^{-5}$$

32% 12.79 (e) $CH_3COOH(aq) + NaOH(aq) \rightarrow NaCH_3COO(aq) + H_2O$

Remember there is a two-fold dilution. More students selected (a) than those who selected the correct response.

68% 12.80 (d) $NH_3 + H_2O \rightleftarrows NH_4^+ + OH^-$

$$K_b = 1.8 \times 10^{-5} = \frac{[NH_4^+][6 \times 10^{-6}]}{[NH_3]} \quad ; \quad \frac{[NH_4^+]}{[NH_3]} = \frac{1.8 \times 10^{-5}}{6 \times 10^{-6}}$$

12.81 A buffer that consists of a mixture of sodium acetate and acetic acid (K_a = 1.8 x 10^{-5}) has a pH of 5.05. The $[CH_3COO^-]$ / $[CH_3COOH]$ ratio in this buffer is

a. 1:2 b. 1:3 c. 1:4 d. 2:1 e. 3:1

The figure below is to be used for Questions 12.82 and 12.83. The curve represents the titration of a weak base by a strong acid.

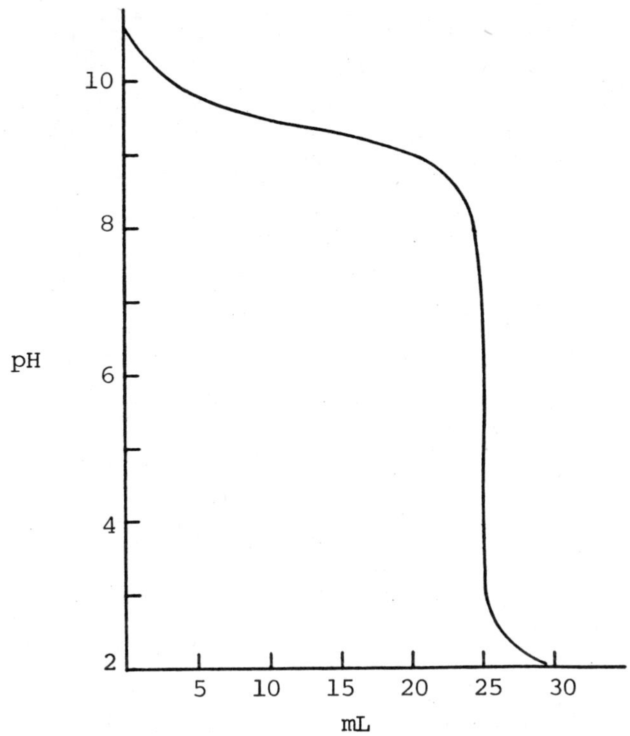

12.82 The equivalence point is represented best by

a. pH = 9-10 b. pH = 7-8 c. pH = 7 d. pH = 5-6 e. pH = 2

12.83 The best indicator for the titration is to be selected from the indicators listed below:

Indicator	pH Color Change Interval
Thymol blue	1.2 - 2.8
Methyl red	4.2 - 6.3
Bromothymol blue	6.2 - 7.6
Phenolphthalein	8.3 -10.0
Alizarin yellow 66	10.0 -12.0

a. Thymol blue b. Methyl red c. Bromothymol blue

d. Phenolphthalein e. Alizarin yellow 66

76% 12.81 (d) $[H_3O^+] = 10^{-pH} = 10^{-5.05} = 8.9 \times 10^{-6}$ M

$$K_a = 1.8 \times 10^{-5} = \frac{[H_3O^+][CH_3COO^-]}{[CH_3COOH]}$$

$$\frac{[CH_3COO^-]}{[CH_3COOH]} = \frac{1.8 \times 10^{-5}}{8.9 \times 10^{-6}}$$

12.82 (d) The equivalence point is the midpoint of the curve with the steep inflection or slope.

12.83 (b) The best indicator is one that will change color around the equivalence point.

The figures below are to be used for Questions 12.84 - 12.87.

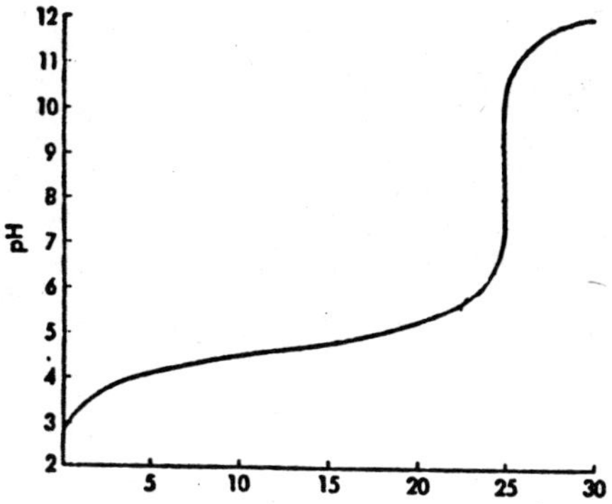

Figure 1

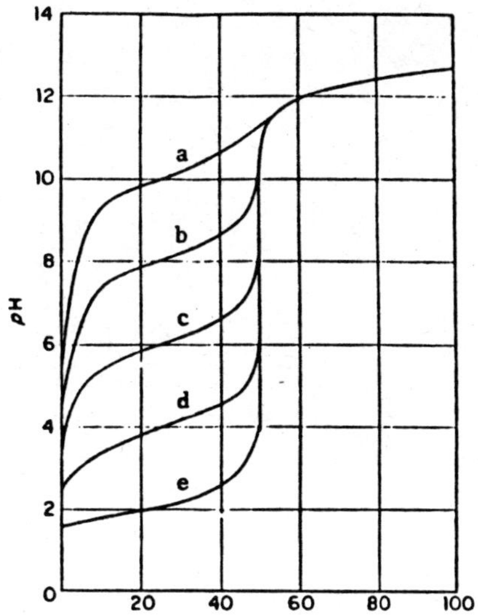

Figure 2

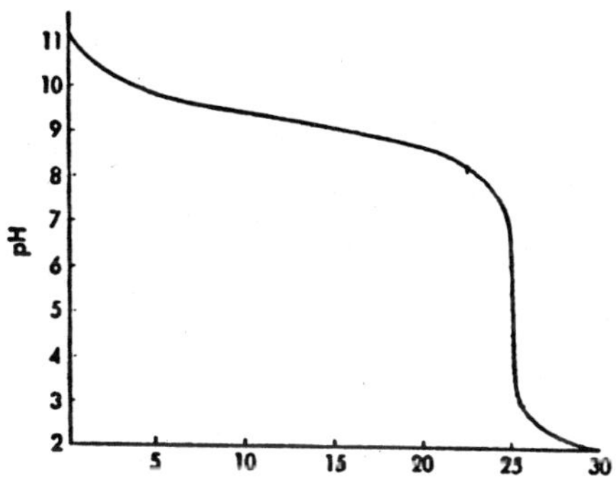

Figure 3

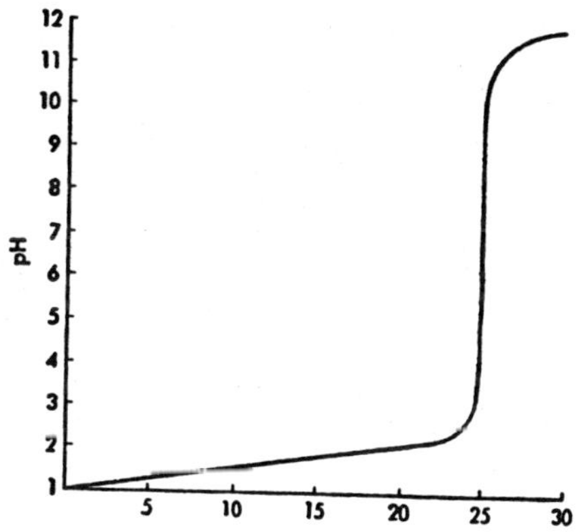

Figure 4

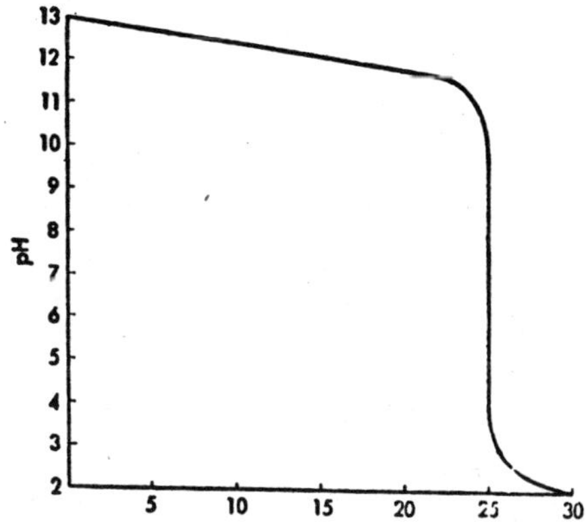

Figure 5

12.84 The titration curve which best represents the titration of a strong acid with a strong base is

a. Figure 1 b. Figure 2, curve d c. Figure 3
d. Figure 4 e. Figure 5

12.85 In Figure 3, the equivalence point is best represented by

a. pH = 2 b. pH = 5-6 c. pH = 7 d. pH = 9-10 e. pH = 11

12.86

Indicator	pH Color Change Interval
Thymol blue	1.2 - 2.8
Methyl red	4.2 - 6.3
Bromothymol blue	6.2 - 7.6
Phenolphthalein	8.3 -10.0
Alizarin yellow 66	10.0 -12.0

The best indicator for the acid-base titration in Figure 1 is

a. Thymol blue b. Methyl red c. Bromothymol blue
d. Phenolphthalein e. Alizarin yellow 66

12.87 The approximate equilibrium constant for the weak acid represented by curve C in Figure 2 is

a. 3×10^{-4} b. 10^{-6} c. 10^{-7} d. 10^{-8} e. 10^{-9}

Questions 12.88 - 12.92 are concerned with the titration of propionic acid.

12.88 A 25.00 mL sample of propionic acid, CH_3CH_2COOH, of unknown concentration was titrated with 0.104 M KOH. The equivalence point was reached when 35.31 mL of base had been added. The concentration of the original propionic acid is

a. 0.0736 M b. 0.128 M c. 0.147 M d. 0.162 M e. 0.295 M

12.89 What is the concentration of the propionate ion at the equivalence point?

a. 0.0531 M b. 0.0609 M c. 0.0914 M d. 0.128 M e. 0.147 M

68% 12.84 (d) A strong acid will have a low pH value and the initial part of the titration curve will be linear.

92% 12.85 (b) The equivalence point is the midpoint of the steep portion of the titration curve.

84% 12.86 (d) Select an indicator that will change color around the equivalence point (~ pH = 9).

36% 12.87 (b) At the 50% titration point [HA] = [A⁻]. $K_a = \frac{[H^+][A^-]}{[HA]}$ and at the 50% titration point $K_a = [H^+]$.

$[H^+] = 10^{-pH} = 10^{-6}$

About 40% of the students selected response (e).

84% 12.88 (c) # mol HA = # mol KOH

$= 0.104 \text{ mol/L} \times 0.03531 \text{ L} = 3.67 \times 10^{-3} \text{ mol}$

$$M_{HA} = \frac{3.67 \times 10^{-3} \text{ mol}}{0.025 \text{ L}} = 0.147 \text{ M}$$

48% 12.89 (b) # mol A^- = # mol KOH = 3.67×10^{-3} mol

$$M_{A^-} = \frac{3.67 \times 10^{-3} \text{ mol}}{0.025 \text{ L} + 0.03531 \text{ L}} = 0.0609 \text{ M}$$

12.90 K_a for CH_3CH_2COOH is 1.34×10^{-5}. What is K_b for $CH_3CH_2COO^-$?

a. 1.34×10^{9} b. 1.34×10^{-5} c. 5.25×10^{-83} d. 7.46×10^{-10}

e. 1.34×10^{-19}

12.91 What is the OH^- concentration for the potassium propionate solution at the equivalence point?

a. 1.5×10^{-9} b. 1.0×10^{-7} c. 6.7×10^{-6} d. 1.1×10^{-5}

e. 4.3×10^{-4}

12.92 The best indicator to use in the titration of propionic acid with KOH is

	Indicator	Transition Range pH	Color Change
a.	Methyl orange	3.1 - 4.4	red to orange
b.	Methyl red	4.2 - 6.3	red to yellow
c.	Bromothymol blue	6.2 - 7.6	yellow to blue
d.	Thymol blue	8.0 - 9.6	yellow to blue
e.	Alizarin yellow	10.0 -12.0	colorless to yellow

The titration curve below represents the titration of a weak acid, HA with a strong base. This curve is for Questions 12.93 and 12.94.

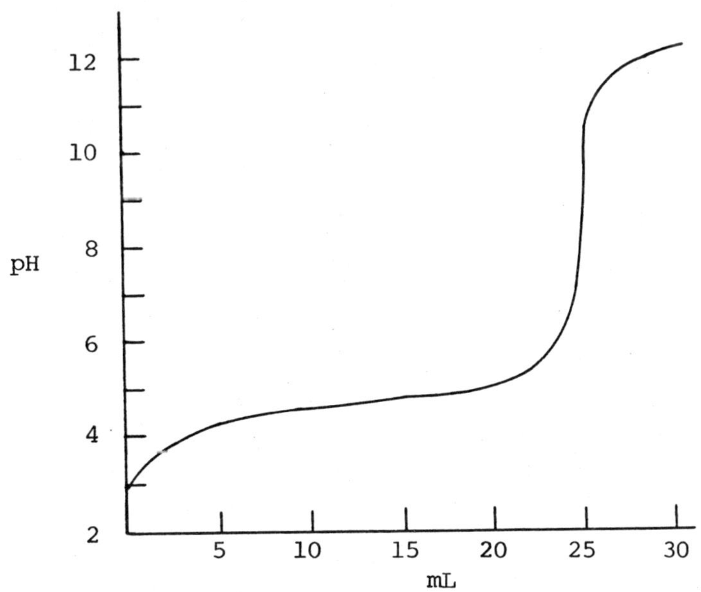

90% 12.90 (d) $K_b = \frac{Kw}{K_a} = \frac{1.00 \times 10^{-14}}{1.34 \times 10^{-5}} = 7.46 \times 10^{-10}$

48% 12.91 (c) $CH_3CH_2COO^- + H_2O \rightleftarrows CH_3CH_2COOH + OH^-$

$0.147 - y \qquad\qquad y \qquad y$

$$K_b = \frac{[CH_3CH_2COOH][OH^-]}{[0.0609 - \cancel{y}]} = \frac{y^2}{0.0609} = 7.46 \times 10^{-10}$$

$y = [OH^-] = 6.7 \times 10^{-6}$ M

64% 12.92 (d) pH at equivalence point equals 14 - pOH = 14.0 - 5.17 = 8.8. Select an indicator whose color changes at the equivalence point.

12.93 The best indicator for the acid-base titration is to be selected from the indicators listed below:

Indicator	pH Color Change Interval
Thymol blue	1.2 - 2.8
Methyl red	4.2 - 6.3
Bromothymol blue	6.2 - 7.6
Phenolphthalein	8.3 -10.0
Alizarin yellow 66	10.0 -12.0

a. Thymol blue b. Methyl red c. Bromothymol blue
d. Phenolphthalein e. Alizarin yellow 66

12.94 The equilibrium constant for the weak acid is
a. 1.3×10^{-3} b. 2.0×10^{-5} c. 1.0×10^{-7}
d. 1.0×10^{-9} e. 1.0×10^{-12}

Hydrolysis

12.95 Which one of the following salts when added to pure water will not change the pH?
a. $NaHCO_3$ b. $NaHSO_4$ c. NaCN d. NH_4NO_3 e. NaBr

12.96 Which of the following when added to water will not change the pH?
a. Na_2CO_3 b. NH_4CH_3COO c. $(NH_4)_3PO_4$ d. NaCN e. NH_4Cl

12.97 One molar solutions of the following three salts: Na_3PO_4, $Al(NO_3)_3$, and NH_4Cl - will result in solutions which are respectively
a. basic, neutral, acidic b. acidic, basic, neutral
c. basic, acidic, acidic d. acidic, basic, neutral
e. acidic, acidic, neutral

80% 12.93 (d) Select the midpoint of the steep portion of the titration curve.

36% 12.94 (b) At the 50% titration point [HA] = [A^-] and for this reaction system [H^+] equals $10^{-4.6}$ = 2.5×10^{-5}.
K_a = [H^+] = 2×10^{-5}

75% 12.95 (e) NaBr is a salt of a strong base and a strong acid. Therefore, the conjugate acid, Na^+, is very weak and the conjugate base, Br^-, is very weak. These ions do react with water.

12.96 (b) All the salts listed react with water. Since NH_3 is as weak a base as CH_3COOH is a weak acid, NH_4^+ and CH_3COO^- will both hydrolyze to the same extent.

12.97 (c) Na_3PO_4 comes from a strong base and a weak acid, so its salt solution will be basic.
$Al(NO_3)_3$ comes from a strong acid and a weak base, so its salt solution will be acidic.
NH_4Cl comes from a weak base and a strong acid, so its salt solution will be acidic.

12.98 Which of the following solutions would be expected to be the most basic?

a. 0.10 M NH_3 b. 0.10 M PO_4^{3-} c. 0.10 M HPO_4^{2-}
d. 0.10 M $N_2H_5^+$ e. 0.10 M $CH_3NH_3^+$

12.99 The two acid dissociation constants for sulfurous acid, H_2SO_3, are 1.2×10^{-2} and 6.2×10^{-8}. The base constant, K_b, or hydrolysis constant for SO_3^{2-} is

a. 2.5×10^{-2} b. 1.6×10^{-7} c. 7.4×10^{-10} d. 8.3×10^{-13}
e. 6.2×10^{-22}

12.100 A 0.1 M solution of ammonia, $K_b = 1.8 \times 10^{-5}$, is used to titrate a 0.1 M HCl solution. At the equivalence point, the pH is approximately

a. 1 b. 5 c. 7 d. 11 e. 13

The information below is to be used for the next two questions. K_1 and K_2 for oxalic acid, $H_2C_2O_4$, are 5.36×10^{-2} and 5.35×10^{-5} respectively.

12.101 What is the equilibrium constant for the reaction

$$C_2O_4^{2-} + H_2O \rightleftarrows HC_2O_4^- + OH^- \ ?$$

a. 5.36×10^{-2} b. 5.3×10^{-5} c. 2.8×10^{-8}
d. 1.9×10^{-10} e. 1.9×10^{-13}

12.102 What is the $[OH^-]$ concentration of a solution which is 0.10 M in $Na_2C_2O_4$?

a. 0.073 b. 2.3×10^{-4} c. 5.3×10^{-5} d. 4.4×10^{-6}
e. 1.4×10^{-7}

55% 12.98 (b) The phosphate ion, PO_4^{3-}, has a lot of negative charge so it will interact very strongly with water.

$$PO_4^{3-} + H_2O \rightleftarrows HPO_4^{2-} + OH^-$$

Because of the negative charge it will of necessity be a stronger base than HPO_4^{2-}. Both $H_2H_5^+$ and $CH_3NH_3^+$ will give acidic solutions.

12.99 (b) $SO_3^{2-} + H_2O \rightleftarrows HSO_3^- + OH^-$

$$K_b = \frac{[HSO_3^-][OH^-]}{[SO_3^{2-}]} = \frac{K_w}{K_2}$$

12.100 (b) At the equivalence point we have an $NH_4Cl(aq)$ solution. The ammonium ion reacts slightly with water

$$NH_4^+ + H_2O \rightleftarrows H_3O^+ + NH_3$$

to give a slightly acidic solution.

80% 12.101 (d) $K_b = \frac{[HC_2O_4^-][OH^-]}{[C_2O_4^{2-}]} = \frac{K_w}{K_2} = \frac{1.0 \times 10^{-14}}{5.35 \times 10^{-5}} = 1.9 \times 10^{-10}$

65% 12.102 (d) $C_2O_4^{2-} + H_2O \rightleftarrows HC_2O_4^- + OH^-$

0.10 - y y y

$$K_b = \frac{y^2}{0.10 - y} = 1.9 \times 10^{-10}$$

$y = [OH^-] = 4.4 \times 10^{-6}$ M

12.103 What is the $[OH^-]$ concentration of a solution which is 0.10 M in NaF (K_a HF is 7.2×10^{-4})?

a. 8.5×10^{-3} b. 1.2×10^{-6} c. 8.3×10^{-9} d. 8.5×10^{-10}

e. 1.4×10^{-11}

Solubility Product

12.104 The solubility product expression for $La_2(CO_3)_3$ is K_{sp} =

a. $[La^{3+}]^3[CO_3^{2-}]^2$ b. $[La^{2+}]^3[CO_3^{2-}]^2$ c. $[La^{3+}]^2[CO_3^{2-}]^3$

d. $[2La^{3+}]^2[3CO_3^{2-}]^3$ e. $[2La^{3+}][3CO_3^{2-}]$

12.105 The solubility product expression for $Fe(OH)_3$ is

a. $[Fe^{3+}][3OH^-]^3$ b. $[Fe^{2+}][OH^-]^3$ c. $[Fe^{3+}][OH^-]^3$

d. $[Fe][OH]^3$ e. $[Fe^{3+}][OH^-]$

12.106 The solubility product expression for $Tb_3(PO_4)_4$ is K_{sp} =

a. $[Tb^{4+}]^3[PO_4^{3-}]^4$ b. $[3Tb^{3+}][4PO_4^{3-}]$ c. $[3Tb^{3+}]^3[4PO_4^{3-}]^4$

d. $[Tb^{3+}]^4[PO_4^{4-}]^3$ e. $[Tb^{2+}]^3[PO_4^{3-}]^2$

12.103 (b) $F^- + H_2O \rightleftarrows HF + OH^-$

0.10-y y y

$$K_b \frac{[HF][OH^-]}{[F^-]} = \frac{Kw}{K_a} = \frac{1.0 \times 10^{-14}}{7.2 \times 10^{-4}} = 1.4 \times 10^{-11}$$

$$y^2 = 0.10 \times 1.4 \times 10^{-11}$$
$$y = [OH^-] = 1.2 \times 10^{-6}\ M$$

92% 12.104 (c) The solubility product constant is equal to the product of the concentration of the ions in a saturated solution, each raised to a power equal to its coefficient in the balanced equation.

$$La_2(CO_3)_3(s) \rightleftarrows 2\ La^{3+}(aq) + 3\ CO_3^{2-}(aq)$$

12.105 (c) See the solution to question 12.104.

$$Fe(OH)_3(s) \rightleftarrows Fe^{3+}(aq) + 3\ OH^-(aq)$$

12.106 (a) See the solution to question 12.104.

$$Tb_3(PO_4)_4(s) \rightleftarrows 3\ Tb^{4+}(aq) + 4\ PO_4^{3-}(aq)$$

The data below are for Questions 12.107 and 12.108

	Salt	K_{sp}
a.	$CaCO_3$	4.8×10^{-9}
b.	PbI_2	1.1×10^{-9}
c.	AgBr	5×10^{-13}
d.	$Fe(OH)_2$	8×10^{-16}
e.	$Co(OH)_2$	2.5×10^{-16}

12.107 Which salt is the least soluble in water? a. b. c. d. e.

12.108 Which salt is the most soluble in water? a. b. c. d. e.

The following information is to be used for Questions 12.109 and 12.110.

	Salt	K_{sp}
a.	$Ni(OH)_2$	6.5×10^{-18}
b.	FeS	5×10^{-18}
c.	$Co(OH)_2$	2.5×10^{-16}
d.	$BaCO_3$	5.1×10^{-9}
e.	PbI_2	1.1×10^{-9}

12.109 Which salt is the most soluble in water? a. b. c. d. e.

12.110 Which salt is the least soluble in water? a. b. c. d. e.

12.107 (c) Since there are both 1:1 and 1:2 salts, one must do some calculations. Find the smallest K_{sp} for both the 1:1 and 1:2 salts. AgBr is the most insoluble 1:1 salt and $Co(OH)_2$ is the most insoluble 1:2 salt.

$AgBr \rightleftarrows Ag^+ + Br^-$; $[Ag^+][Br^-] = 5 \times 10^{-13}$
$[Ag^+]^2 = 5 \times 10^{-13}$; $[Ag^+]$ = molar solubility = 7×10^{-7} M
$Co(OH)_2 \rightleftarrows Co^{2+} + 2\ OH^-$; $4\ y^3 = 2.5 \times 10^{-16}$
$y = 4 \times 10^{-6}$ M = $[Co^{2+}]$ = molar solubility.

12.108 (b) Select the 1:1 electrolyte which has the largest K_{sp} value ($CaCO_3$) and the 1:2 electrolyte with the largest K_{sp} value (PbI_2). Since both these salts have K_{sp} values of 10^{-9}, PbI_2 must be more soluble because its molar solubility involves taking the cube root of the K_{sp} whereas $CaCO_3$ involves the square root of the K_{sp} value.

68% 12.109 (e) We need to be careful because there are both 1:1 and 1:2 salts. We cannot just look at the largest K_{sp} value. The K_{sp} value for $BaCO_3$ is slightly larger than that for PbI_2, but the molar solubility of $BaCO_3$ is proportional to $K_{sp}^{1/2}$ whereas the molar solubility of PbI_2 is proportional to the $K_{sp}^{1/3}$.

88% 12.110 (b) See problem 12.109 above. Look for the smallest K_{sp} values. $Ni(OH)_2$ and FeS have the smallest K_{sp} values. Infact $(5 \times 10^{-18})^{1/2}$ is much smaller than $(\frac{6.5 \times 10^{-18}}{4})^{1/3}$.

12.111

	K_{sp}
AgCl	1.0×10^{-10}
Ag_2CO_3	6.3×10^{-12}
Ag_3PO_4	2.0×10^{-21}

From a consideration of the data above, the order for the molar solubility for the salts is

a. $AgCl > Ag_2CO_3 > Ag_3PO_4$ b. $Ag_2CO_3 > Ag_3PO_4 > AgCl$
c. $Ag_2CO_3 > AgCl > Ag_3PO_4$ d. $Ag_3PO_4 > Ag_2CO_3 > AgCl$
e. $AgCl > Ag_3PO_4 > Ag_2CO_3$

12.112 The $[OH^-]$ of a saturated solution of $Ca(OH)_2$ is 0.025 M. What is K_{sp} for calcium hydroxide?

a. 6.3×10^{-4} b. 1.6×10^{-5} c. 7.8×10^{-6}
d. 1.6×10^{-25} e. 2.3×10^{-26}

12.113 If the pH of a solution of $Ba(OH)_2$ is 10.00 at 25°C, then the barium ion concentration is

a. 1.0×10^{-4} b. 2.0×10^{-4} c. 5.0×10^{-5} d. 1.0×10^{-10}
e. 5.0×10^{-11}

12.114 The K_{sp} for $BaSO_4$ is 1.1×10^{-10}. The molar solubility of $BaSO_4$ is

a. 1.1×10^{-10} M b. 1.8×10^{-10} M c. 1.0×10^{-5} M
d. 1.3×10^{-5} M e. 4.6×10^{-5} M

12.111 (c) Let y = the molar solubility.

AgCl; $[y][y] = 1.0 \times 10^{-10}$; $y = [Ag^+] = 1.0 \times 10^{-5}$ M

Ag_2CO_3; $[2y]^2[y] = 6.3 \times 10^{-12}$; $y = [CO_3^{2-}] = 1.2 \times 10^{-4}$M

Ag_3PO_4; $[3y]^3[y] = 2.0 \times 10^{-21}$; $y = [PO_4^{3-}] = 2.9 \times 10^{-6}$M

88% 12.112 (c) $Ca(OH)_2(s) \rightleftarrows Ca^{2+}(aq) + 2\ OH^-(aq)$

0.025/2 0.025

$K_{sp} = [0.025/2][0.025]^2 = 7.8 \times 10^{-6}$

12.113 (c) $[H_3O^+] = 1.0 \times 10^{-10}$

$[OH^-] = 1.0 \times 10^{-4}$

$$[Ba^{2+}] = \frac{1[Ba^{2+}]}{2[OH^-]} \times 1.0 \times 10^{-4}\ M\ OH^- = 5.0 \times 10^{-5}\ M$$

96% 12.114 (c) $BaSO_4 \rightleftarrows Ba^{2+} + SO_4^{2-}$

$[Ba^{2+}][SO_4^{2-}] = 1.1 \times 10^{-10}$

$[y][y] = 1.1 \times 10^{-10}$; $y^2 = 1.1 \times 10^{-10}$

$[Ba^{2+}] = y =$ molar solubility $= 1.0 \times 10^{-5}$

12.115 One hundred milliters of a saturated solution of calcium sulfate contains 0.209 grams of calcium sulfate at 30°C. What is the solubility product of calcium sulfate? (Molecular weight of calcium sulfate is 136.)

a. 1.0×10^{-2} to 2.0×10^{-2} b. 3.0×10^{-3} to 3.5×10^{-3}
c. 1.5×10^{-1} to 2.5×10^{-1} d. 2.0×10^{-4} to 2.5×10^{-4}
e. None of the above.

12.116 The molar solubility for lead fluoride is 2.1×10^{-3} at 25°C. Calculate the solubility product constant for lead fluoride.

a. 4.4×10^{-6} b. 4.2×10^{-7} c. 9.3×10^{-7}
d. 3.7×10^{-8} e. 9.3×10^{-9}

12.117 The $[Mg^{2+}]$ of a saturated solution of $Mg(OH)_2$ is 1.45×10^{-4} M. What is the K_{sp} of $Mg(OH)_2$?

a. 6.1×10^{-12} b. 1.2×10^{-11} c. 2.4×10^{-11}
d. 9.8×10^{-11} e. None of these.

12.118 What is the molar solubility of $SrSO_4$ in 0.10 M Na_2SO_4? (K_{sp} - for $SrSO_4$ = 2.5×10^{-7}.)

a. 2.5×10^{-7} b. 2.5×10^{-6} c. 5.0×10^{-4}
d. 5.0×10^{-3} e. 2.5×10^{-3}

12.119 The molar solubility of MgF_2 in a 0.20 M KF solution is (K_{sp} for MgF_2 = 8.0×10^{-8})

a. 0.20 M b. 2.0×10^{-6} M c. 5.0×10^{-7} M d. 2.0×10^{-7} M
e. 8.0×10^{-8} M

44% 12.115 (d) $CaSO_4(s) \rightleftarrows Ca^{2+}(aq) + SO_4^{2-}(aq)$

$S_{CaSO_4} = S_{Ca^{2+}} = S_{SO_4^{2-}}$

$S_{CaSO_4} = 2.09 \text{ g/L} \times \frac{1 \text{ mole}}{136 \text{ g}} = 0.0154 \text{ M}$

$K_{sp} = [Ca^{2+}][SO_4^{2-}] = [0.0154]^2 = 2.4 \times 10^{-4}$

48% 12.116 (d) $PbF_2 \rightleftarrows Pb^{2+} + 2 F^-$

$[Pb^{2+}] = 2.1 \times 10^{-3}$; $[F^-] = 2 \times 2.1 \times 10^{-3}$

$[2.1 \times 10^{-3}][4.2 \times 10^{-3}]^2 = K_{sp} = 3.7 \times 10^{-8}$

12.117 (b) $Mg(OH)_2(s) \rightleftarrows Mg^{2+}(aq) + 2 OH^-(aq)$

$1.45 \times 10^{-4}\text{M} \quad 2.90 \times 10^{-4}\text{M}$

$K_{sp} = [1.45 \times 10^{-4}][2.90 \times 10^{-4}]^2 = 1.2 \times 10^{-11}$

12.118 (b) $SrSO_4(s) \rightleftarrows Sr^{2+}(aq) + SO_4^{2-}(aq)$

$y \quad 0.10 + y$

$[y][0.10 + \cancel{y}] = 2.5 \times 10^{-7}$

$y = [Sr^{2+}] = \text{molar solubility} = 2.5 \times 10^{-6} \text{ M}$

12.119 (b) $MgF_2 \rightleftarrows Mg^{2+} + 2F^-$

$y \quad 0.20 + 2y$

$[y][0.20 + \cancel{2y}]^2 = 8.0 \times 10^{-8}$

Since 2y is small compared to 0.20, let us ignore it.

$$\therefore\ y = [Mg^{2+}] = \text{molar solubility} = \frac{8.0 \times 10^{-8}}{[0.20]^2}$$

$$y = 2.0 \times 10^{-6}$$

Be sure to check as to where the assumption was valid.

$$0.20 + 2(2.0 \times 10^{-6}) = 0.20$$

12.120 The solubility product of CaF_2 is 3.9×10^{-11}. In a solution in which the fluoride concentration is 0.20 M, the equilibrium concentration of $[Ca^{2+}]$ is

a. 3.9×10^{-11} b. 9.8×10^{-11} c. 2.0×10^{-10} d. 2.4×10^{-10}
e. 9.8×10^{-10}

12.121 A solution is 0.010 M in each of $Pb(NO_3)_2$, $Mn(NO_3)_2$ and $Zn(NO_3)_2$. Solid NaOH is added until the $[OH^-]$ of the solution is 4.0×10^{-7} M. K_{sp} = 2.8×10^{-16} for $Pb(OH)_2$, 4.5×10^{-14} for $Mn(OH)_2$, and 4.5×10^{-17} for $Zn(OH)_2$. Which is true?

a. There will be no precipitate formed.
b. Only $Zn(OH)_2$ will precipitate.
c. Only $Mn(OH)_2$ will precipitate.
d. Only $Zn(OH)_2$ and $Pb(OH)_2$ will precipitate.
e. All three hydroxides will precipitate.

12.122 A solution is 0.010 M in each of $Pb(NO_3)_2$, $Mn(NO_3)_2$, and $Zn(NO_3)_2$. Solid NaOH is added until the $[OH^-]$ of the solution is 3.0×10^{-6} M. K_{sp} = 2.8×10^{-16} for $Pb(OH)_2$, 4.5×10^{-14} for $Mn(OH)_2$, and 4.5×10^{-17} for $Zn(OH)_2$. Which is true?

a. There will be no precipitate formed.
b. Only $Zn(OH)_2$ will precipitate.
c. Only $Mn(OH)_2$ will precipitate.
d. Only $Zn(OH)_2$ and $Pb(OH)_2$ will precipitate.
e. All three hydroxides will precipitate.

12.120 (e) $CaF_2(s) \rightarrow Ca^{2+}(aq) + 2\ F^-(aq)$

$$y \qquad 0.20 + 2y$$

↑ ignore

$[y][0.20]^2 = 3.9 \times 10^{-11}$

$y = 9.8 \times 10^{-10}$ = molar solubility of CaF_2

12.121 (d) First, calculate a trial product.

$[M^{2+}][OH^-]^2 = K_{sp}$

$(0.010)(4.0 \times 10^{-7})^2 = 1.6 \times 10^{-15}$

Since $1.6 \times 10^{-15} > 2.8 \times 10^{-16}$ and 4.5×10^{-17}, both $Zn(OH)_2$ and $Pb(OH)_2$ will precipitate.

12.122 (e) First, calcualte a trial product.

$[M^{2+}][OH^-]^2 = K_{sp}$

$(0.010)(3.0 \times 10^{-6})^2 = 9.0 \times 10^{-14}$

Since $9.0 \times 10^{-14} > 4.5 \times 10^{-14}$, 2.8×10^{-16} and 4.5×10^{-17}, all three hydroxides will precipitate.

12.123 Given the following information for four sulfides:

Sulfide	K_{sp}
CdS	3.6×10^{-29}
ZnS	4.9×10^{-18}
CuS	8.7×10^{-36}
PbS	8.4×10^{-28}

To saturated 1.0 L solutions of all the sulfides above, 100 mL of 0.01 M sodium sulfide is added. Which one of the following ions exists in lowest concentration at 25°C?

a. Cd^{2+} b. Zn^{2+} c. Cu^{2+} d. Pb^{2+} e. S^{2-}

12.124 The best way to ensure complete precipitation of ZnS from a saturated H_2S solution is to

a. add more H_2S. b. add a strong acid. c. add a weak acid.
d. increase the pH. e. heat the solution.

12.125 From a consideration of the reactions

$$Ag^+(aq) + 2NH_3(aq) \rightleftarrows Ag(NH_3)_2^+(aq) \qquad K = 1.8 \times 10^7$$
$$Ag^+(aq) + Cl^-(aq) \rightleftarrows AgCl(s) \qquad K = 5.6 \times 10^9$$

calculate the equilibrium constant for

$$AgCl(s) + 2NH_3(aq) \rightleftarrows Ag(NH_3)_2^+(aq) + Cl^-(aq)$$

a. 1.0×10^{-17} b. 3.2×10^{-3} c. 1.0×10^{17} d. 1.0×10^3
e. 3.1×10^{16}

80% 12.123 (c) This question is just asking which sulfide is the most insoluble. Since all the sulfides have formed 1:1 salts, then we only need to find the smallest K_{sp} value.

12.124 (d) Increase the sulfide ion concentration. The sulfide ion concentration can be increased by adding base (increasing the pH).

$$H_2S + 2\ OH^- \rightleftarrows 2\ H_2O + S^{2-}$$

50% 12.125 (b)

Reaction	K
$AgCl(s) \rightleftarrows Ag^+(aq) + Cl^-(aq)$	$1/5.6 \times 10^9$
$Ag^+(aq) + 2NH_3(aq) \rightleftarrows Ag(NH_3)_2^+(aq)$	1.8×10^7
$AgCl(s) + 2NH_3(aq) \rightleftarrows Ag(NH_3)_2^+(aq) + Cl^-(aq)$	3.2×10^{-3}

The data below are for Questions 12.126 and 12.127.

$H_2S + H_2O \rightleftarrows H_3O^+ + HS^-$ $K_1 = 1.0 \times 10^{-7}$

$HS^- + H_2O \rightleftarrows H_3O^+ + S^{2-}$ $K_2 = 1.3 \times 10^{-13}$

12.126 A saturated H_2S solution is 0.10 M and it is buffered at a pH of 3.00. What is the concentration of S^{2-} in this solution?

a. 0.10 M b. 1.3×10^{-15} M c. 3.2×10^{-16} M
d. 1.3×10^{-18} M e. 6.5×10^{-19} M

12.127 The sulfide ion concentration in a solution which is 0.10 M with respect to HCl and is saturated with H_2S at 0.10 M concentration is

a. $1. \times 10^{-8}$ M b. 1.3×10^{-14} M c. 1.3×10^{-19} M
d. 1.3×10^{-20} M e. 1.3×10^{-21} M

76% 12.126 (b) $\underbrace{H_2S}_{0.10 - y} + 2\ H_2O \rightleftarrows \underbrace{2\ H_3O^+}_{1.00 \times 10^{-3} + y} + \underbrace{S^{2-}}_{y}$ $K_1K_2 = 1.2 \times 10^{-20}$

$$\frac{[H_3O^+]^2[S^{2-}]}{[H_2S]} = 1.3 \times 10^{-20}$$

$$\frac{[1.00 \times 10^{-3}]^2[y]}{[0.10]} = 1.3 \times 10^{-20}$$

$$y = [S^{2-}] = 1.3 \times 10^{-15}\ M$$

72% 12.127 (c) $\dfrac{[H_3O^+]^2[S^{2-}]}{[H_2S]} = 1.3 \times 10^{-20}$

$$\frac{[0.10]^2[y]}{[0.10]} = 1.3 \times 10^{-20}$$

$$y = [S^{2-}] = 1.3 \times 10^{-19}\ M$$

13

THERMODYNAMICS

Thermochemistry

13.1 The amount of heat needed to raise the temperature of 1.00 g of iron 1°C is called its

a. entropy b. enthalpy c. specific heat d. molar heat

e. molar heat capacity

13.2 The quantity of heat needed to raise the temperature of 46.1 g of ethanol, C_2H_5OH, 1°C is called its

a. specific heat b. molar enthalpy c. molar entropy

d. molar heat capacity e. specific heat per mole

13.3 Given the following specific heats of metals:

Metal	Specific Heat (J/g°C)
Copper	0.385
Magnesium	1.02
Mercury	0.138
Silver	0.237
Lead	0.129

If the same amount of heat is added to 50.0 g samples of each of the metals which are all at the same temperature, which metal will reach the highest temperature?

a. Copper b. Magnesium c. Mercury d. Silver e. Lead

13.4 What is the specific heat of benzene if 3450 J of heat is added to 150. g sample of benzene and its temperature increases from 22.5 °C to 35.8 °C.

a. 0.64 J/g·°C b. 1.02 J/g·°C c. 1.73 J/g·°C d. 2.61 J/g·°C

e. 23.0 J/g·°C

13.1 (c) This is the definition of specific heat.

13.2 (d) 46.1 g of ethanol is the mass of one mole of ethanol. The molar heat capacity is the amount of heat to raise the temperature of one mole of substance 1°C.

66% 13.3 (e) Which metal has the smallest specific heat?

13.4 (c) $\text{sp. heat} = \dfrac{3450\ \text{J}}{150.\ \text{g} \times (35.8 - 22.5)^\circ\text{C}} = 1.73\ \dfrac{\text{J}}{\text{g}\cdot{}^\circ\text{C}}$

13.5 The specific heat of lead is 0.129 J/g·°C. If 74.0 J of heat is added to a 125 g piece of lead at 23.0°C, what is the final temperature of the lead?

a. 18.4 °C b. 23.0 °C c. 26.4 °C d. 27.6 °C e. 28.4 °C

13.6 What is the <u>molar</u> heat of combustion of ethanol, C_2H_5OH, if combustion of 2.00 g of ethanol causes a temperature rise of 3.29 °C in a bomb calorimeter that has a heat capacity of 17.9 kJ/°C? The molecular weight of C_2H_5OH = 46.1.

a. -2.56 kJ/mol b. -29.4 kJ/mol c. -58.9 kJ/mol d. -1360 kJ/mol
e. -2720 kJ/mol

13.7 500 cm^3 of 1.0 M NaOH(aq) is added to 500 cm^3 of 1.0 M HCl(aq) and the solution is quickly stirred. The rise in temperature (T_1) is measured. The experiment is repeated using 100 cm^3 of each solution and the rise in temperature (T_2) is measured. It is found that

a. T_2 is five times as large at T_1. b. T_1 is five times as large as T_2.
c. T_2 is greater than T_1. d. T_2 is equal to T_1 e. $\Delta T_1 = 5\Delta T_2$.

13.8 When 0.050 mol of HCl is reacted with 0.050 mol of NaOH in 50.0 mL of water, the temperature of the water increases by 13.7°C. Calculate the molar heat of neutralization

$$HCl(aq) + NaOH(aq) \rightarrow NaCl(aq) + H_2O(\ell)$$

The heat capacity of the system is 209.2 J/°C.

a. -57,300 J b. -2,870 J c. +2,870 J d. +57,300 J e. +115,000 J

13.9 Calculate the amount of heat evolved in the complete oxidation of 8.17 g of Al at 25°C and 1 atm pressure. ΔH_f° for $Al_2O_3(s)$ is -1676 kJ/mol.

$$4Al(s) + 3O_2(g) \rightarrow 2Al_2O_3(s)$$

a. 101 kJ b. 127 kJ c. 203 kJ d. 237 kJ e. 254 kJ

13.5 (d) $74.0 \text{ J} = 125 \text{ g} \times 0.129 \frac{\text{J}}{\text{g}\cdot{}^\circ\text{C}} \times (t_f - 23.0)^\circ\text{C}$

$$16.13\, t_f = 370.9 + 74.0 = 444.9$$

$$t_f = 27.6^\circ\text{C}$$

13.6 (d) $$\frac{-17.9 \frac{\text{kJ}}{^\circ\text{C}} \times 3.29^\circ\text{C}}{2.00 \text{ g}} \times \frac{46.1 \text{ g}}{1 \text{ mol}} = -1360 \frac{\text{kJ}}{\text{mol}}$$

55% 13.7 (d) The concentrations of the solutions haven't changed and if all the heat that is given off is used to increase the temperature of the solution then the temperatures will be the same.

13.8 (a) $$\frac{-209.2 \frac{\text{J}}{^\circ\text{C}} \times 13.7\ ^\circ\text{C}}{0.050 \text{ mol}} = -57300 \frac{\text{J}}{\text{mol}}$$

13.9 (e) $$\#\text{ mol } Al_2O_3 = \frac{2\ Al_2O_3}{4\ Al} \times 8.17 \text{ g} \times \frac{1 \text{ mol Al}}{26.98 \text{ g Al}}$$

$$= 0.151$$

$$\Delta H^\circ = -\ 1676 \frac{\text{kJ}}{\text{mol}} \times 0.151 \text{ mol} = -254 \text{ kJ}$$

State Functions and Their Signs

13.10 The purpose of thermodynamics is to learn about a reaction's

a. pathway b. mechanism c. stoichiometry d. rate
e. feasibility

13.11 The functions listed below which are state functions are

1. S 2. H 3. q 4. w

a. 1 only b. 1 and 2 only c. 2 only d. 1, 2, 3, and 4
e. 3 and 4 only

Questions 13.12 - 13.15 consist of five lettered headings followed by a list of numbered phrases. For each numbered phrase, select the one heading which is most closely related to it. Each heading may be used once, more than one, or not at all in each group.

a. ΔS b. ΔH c. ΔE d. ΔG e. $T\Delta S$

13.12 Measurement of heat at constant volume.

13.13 The thermodynamic quantity which describes the change in randomness of a chemical reaction.

13.14 Is essentially equal to the change in enthalpy when both the reactants and products are solids.

13.15 Is equal to ΔH at equilibrium.

13.16 Which of the following must have a negative value for an exothermic reaction?

a. Enthalpy change b. Free energy change c. Entropy change
d. Work e. $P\Delta V$

13.10 (e) The study of kinetics is concerned with the pathway (mechanism) and rate whereas thermodynamics is concerned with whether the reaction is possible.

13.11 (b) Both heat and work depend upon the pathway. Therefore, they are not state functions.

13.12 (c) $\Delta E = q - w$

At constant volume, no work is done so $\Delta E = q_v$.

13.13 (a) Entropy (the amount of disorder). We could probably argue that (e) is a correct answer too.

13.14 (c) ΔE is approximately equal to ΔH when there is essentially no change in volume ($\Delta n_{gas} = 0$).

13.15 (e) $\Delta G = \Delta H - T\Delta S$

At equilibrium, there is no driving force so $\Delta G = 0$.

$0 = \Delta H - T\Delta S$ and $\Delta H = T\Delta S$.

64% 13.16 (a) $\Delta G = \Delta H - T\Delta S$

ΔG can be positive even when ΔH is negative if ΔS is negative and the temperature is sufficiently high.

The next four questions pertain to the material given below. Each question below consists of the five lettered headings below. Select the one heading which is most closely related to the question. A heading can be used more than once.

a. $Br_2(\ell) \rightarrow Br_2(g)$ b. $Br_2(g) \rightarrow 2\ Br(g)$ c. $Br(g) + e \rightarrow Br^-(g)$
d. $K^+(g) + Br^-(g) \rightarrow KBr(s)$ e. $K(s) + 1/2\ Br_2(\ell) \rightarrow KBr(s)$

13.17 Process which represents the crystal lattice energy.

13.18 Process which represents the bond dissociation energy.

13.19 Process which respresents the most exothermic step.

13.20 Process which represents the heat of formation.

13.21 The enthalpy change for the following processes would be expected to be negative for

1. $Na^+(g) + e \rightarrow Na(g)$ 2. $F(g) + e \rightarrow F^-(g)$
3. $Na^+(g) + F^-(g) \rightarrow NaF(s)$ 4. $H_2O(\ell) \rightarrow H_2O(g)$

a. 1 and 2 only b. 2 and 4 only c. 1, 2 and 3 only d. 4 only
e. 1 and 3 only

13.22 Of the following processes

1. $H_2(g) \rightarrow 2H(g)$ 2. $H_2O(\ell) \rightarrow H_2O(s)$ 3. $CO_2(s) \rightarrow CO_2(g)$

the one(s) which is(are) exothermic ($\Delta H = -$) is(are)

a. 1 only b. 2 only c. 3 only d. 1 and 2 only e. 2 and 3 only

13.23 Which one of the following processes has the sign of its enthalpy change different from the rest?

a. $2\ CH_3(g) \rightarrow C_2H_6(g)$ b. $Cl_2(g) \rightarrow 2\ Cl(g)$
c. $Cl(g) + e \rightarrow Cl^-(g)$ d. $2\ Fe(s) + 3/2\ O_2(g) \rightarrow Fe_2O_3(s)$
e. $Na^+(g) + e \rightarrow Na(g)$

80% 13.17 (d) The crystal lattice energy is the energy given off when a mole of +1 ions at infinite distance from a mole of -1 ions are allowed to come together to form one mole of a crystalline solid.

92% 13.18 (b) The bond dissociation energy is the energy required to break a bond in a gaseous molecule to form gaseous species (atoms).

60% 13.19 (d) In the processes that are depicted, steps (c), (d) and (e) are all exothermic. Step (d), the crystal lattice energy is a very exothermic process. Ionic bonds are being formed and a phase change is occurring.

13.20 (e) The standard molar enthalpy of formation is the amount of heat involved in a reaction in which one mole of a product in its standard state is formed from its elements in their standard states.

60% 13.21 (c)
1. The reverse of ionization energy is an exothermic process.
2. The electron affinity for a nonmetal is an exothermic process.
3. The crystal lattice energy is highly exothermic.
4. The heat of vaporization is always endothermic.

64% 13.22 (b)
1. Bond breaking is an endothermic process.
2. Melting of ice requires heat, so the reverse of this process would be exothermic.
3. Sublimation is an endothermic process.

48% 13.23 ;(b) This is an endothermic process because bond breaking is involved. Reactions between elements (Fe and O_2) are usually always exothermic since bonds between unlike atoms tend to be stronger than bonds between like atoms.

13.24 Which one of the following processes has the sign of its enthalpy change different from the rest?

a. $Al_2O_3(s) \rightarrow 2Al(s) + 3/2\ O_2(g)$ b. $H_2O(s) \rightarrow H_2O(\ell)$

c. $Cl_2(g) \rightarrow 2\ Cl(g)$ d. $Cl(g) + e \rightarrow Cl^-(g)$

e. $H_2O(\ell) \rightarrow H_2(g) + 1/2\ O_2(g)$

13.25 Which one of the following processes is exothermic?

a. $SO_3(g) \rightarrow SO_2(g) + 1/2\ O_2(g)$

b. $Mg^+(g) \rightarrow Mg^{2+}(g) + e$

c. $NH_4Cl(s) \rightarrow NH_3(g) + HCl(g)$

d. $CO_2(s) \rightarrow CO_2(g)$

e. $2F(g) \rightarrow F_2(g)$

13.26 This question is concerned with the heat change when ammonia is formed from its elements

$$N_2(g) + 3H_2(g) \rightarrow 2NH_3(g) \quad \Delta H = -92\ kJ$$

Therefore, 92 kJ is the quantity of heat which is

a. lost to the surroundings when 1 mole of hydrogen is consumed.

b. lost to the surroundings when 1 mole of ammonia is formed.

c. lost to the surroundings when 2 moles of ammonia are formed.

d. gained from the surroundings when 1 mole of ammonia is formed.

e. gained from the surroundings when 2 moles of ammonia are formed.

13.27 ΔH and ΔE are nearly the same in all the following processes EXCEPT

a. $H_2(g) + I_2(g) \rightarrow 2HI(g)$

b. $H^+(aq) + Cl^-(aq) + Na^+(aq) + OH^-(aq) \rightarrow H_2O(aq) + Na^+(aq) + Cl^-(aq)$

c. $COCl_2(g) \rightarrow CO(g) + Cl_2(g)$

d. $CO(g) + H_2O(g) \rightarrow H_2(g) + CO_2(g)$

e. $Hg(s) \rightarrow Hg(\ell)$

56% 13.24 (d) a. Endothermic; it usually takes energy to form elements from a compound.

b. An endothermic phase change.

c. Bond breaking (endothermic).

d. Electron affinity (usually exothermic for the first electron).

e. See a. above.

60% 13.25 (e) The formation of a chemical bond from atoms is always an exothermic process.

96% 13.26 (c) The reaction is exothermic and the heat lost is for the thermochemical equation that is written.

13.27 (c) $\Delta H = \Delta E$ when Δn_{gas} for the reaction is equal to zero. $\Delta n_{gas} = +1$ for reaction (c).

Enthalpy of Formation

13.28 Of the species listed below, the only one which has a standard enthalpy of formation value of zero is

a. Li(g) b. N(g) c. O(g) d. F(g) e. Ne(g)

13.29 Of the species listed below, the only one which has a standard enthalpy of formation value of zero is

a. Na(ℓ) b. Hg(ℓ) c. Cl_2(ℓ) d. H_2O(ℓ) e. Ne(ℓ)

13.30 For which of the following substances is ΔH_f° equal to 0?

a. F_2(g) b. C(g) c. CO(g) d. Na(g) e. He(ℓ)

13.31 Of the species listed below, the only one which has a standard enthalpy of formation value of zero is

a. O(g) b. O_2(ℓ) c. O_3(g) d. NH_3(g) e. Na(s)

13.32 The standard heat of formation of sodium bromide, NaBr(s), corresponds to the reaction

a. Na(g) + Br(ℓ) → NaBr(s)
b. Na(g) + Br(g) → NaBr(g)
c. 2Na(s) + Br_2(ℓ) → 2NaBr(s)
d. Na(s) + Br(ℓ) → NaBr(s)
e. Na(s) + 1/2 Br_2(ℓ) → NaBr(s)

13.33 Which of the following is the correct thermochemical equation which represents the standard enthalpy of formation of CO(g)?

a. C(g) + 1/2 O_2(g) → CO(g)
b. C(s) + 1/2 O_2(g) → CO(g)
c. C(g) + O(g) → CO(g)
d. CO(g) → C(s) + 1/2 O_2(g)
e. 1/2 C(s) + 1/2 CO_2(g) → CO(g)

88% 13.28 (e) Li(s), $N_2(g)$, $O_2(g)$ and $F_2(g)$ have values of $\Delta H_f^\circ = 0$ at 25°C.

76% 13.29 (b) Which element is in its most stable form at 25°C?

92% 13.30 (a) Which element is in its most stable form at 25°C? Compounds do not have ΔH_f° values of zero.

96% 13.31 (e) See the answer to question 13.30 above.

13.32 (e) Na is a solid at 25°C and Br_2 is a diatomic molecule and a liquid at 25°C. The standard heat of formation is the change in heat for producing one mole of product from the reactants in their most stable form at 25°C.

54% 13.33 (b) The standard enthalpy of formation of a product involves the reaction of the elements in their most stable states ($\Delta H_f^\circ = 0$ for the elements) to form the product. Response (a) is not the answer because carbon is a solid at room temperature not a gas.

13.34 The chemical reaction below which represents the standard ΔH_f° is

a. $3Li(s) + 1/2\ N_2(g) \rightarrow Li_3N(s)$ b. $3Li(g) + 1/2\ N_2(g) \rightarrow Li_3N(s)$

c. $3Li(g) + N(g) \rightarrow Li_3N(s)$ d. $3Li(g) + N(g) \rightarrow Li_3N(g)$

e. $3Li(s) + N(g) \rightarrow Li_3N(g)$

13.35 For which of the following equations is the enthalpy change at 25°C and 1 atm equal to ΔH_f° (HCOOH(ℓ))?

a. $CO_2(g) + H_2(g) \rightarrow HCOOH(\ell)$

b. $CO(g) + H_2O(\ell) \rightarrow HCOOH(\ell)$

c. $C(s) + 2H(g) + 2O(g) \rightarrow HCOOH(\ell)$

d. $C(g) + 2H(g) + 2O(g) \rightarrow HCOOH(\ell)$

e. $C(s) + H_2(g) + O_2(g) \rightarrow HCOOH(\ell)$

13.36 A 1.00 g sample of pure solid carbon at 25°C and 1 atmosphere pressure was burned in oxygen to give a 3.66 g sample of pure gaseous carbon dioxide. The energy given off in this reaction was 32,800 J. Based on this information, the heat of formation of CO_2 is

a. 32,800 J b. -32,800 J c. 120,000 J d. 394,000 J

e. -394,000 J

13.37 At 25°C when 1.00 g of calcium is burned at constant pressure in an atmosphere of oxygen to give CaO, 3.79 kcal of heat energy is liberated. The standard enthalpy of formation of CaO is ________ kcal/mole.

a. -3.79 x 1 b. 3.79 x 1 c. -3.79 x 40.1 d. 3.79 x 40.1

e. -3.79 x 56.1

13.38 The compound $PCl_3(g)$ can be burned in Cl_2 to give PCl_5. A total of 46.5 kJ of energy is liberated in the process when 0.500 mole of $PCl_3(g)$ is burned. What is $\Delta H^\circ_{formation}$ of PCl_5? ($\Delta H_{formation}$ $PCl_3(g)$ = -306 kJ/mole)

a. -93 kJ/mole b. -213 kJ/mole c. -259 kJ/mole

d. -306 kJ/mole e. -399 kJ/mole

13.34 (a) The enthalpy of formation of a compound is the heat of reaction when the compound is formed from the elements in their standard states. Lithium is a solid and nitrogen a gas at 25°C and 1 atmosphere pressure.

56% 13.35 (e) Carbon is a solid, hydrogen and oxygen are diatomic gases at 25°C.

36% 13.36 (e) $-32{,}800\ \frac{J}{1.00\ gC} \times \frac{12.0\ gC}{1\ mole\ C} = -393{,}600\ \frac{J}{mole}$

$$-32{,}800\ \frac{J}{3.66\ g\ CO_2} \times \frac{44.0\ g\ CO_2}{1\ mole\ CO_2} = -394{,}300\ \frac{J}{mole}$$

13.37 (c) $Ca(s) + 1/2\ O_2(g) \rightarrow CaO(s)$

To form one mole of CaO, 40.1 g of Ca needs to be burned.

$$-3.79\ \frac{kcal}{g} \times \frac{40.1\ g}{1\ mol\ Ca}$$

$\Delta H^\circ_f = -152$ kcal/mol

35% 13.38 (e) $PCl_3(g) + Cl_2(g) \rightarrow PCl_5(g)$

$\Delta H_{rxn} = 2(-46.5\ kJ) = -93.0\ kJ$

$-93.0 = \Sigma\Delta H^\circ(prod) - \Sigma\Delta H^\circ(react)$

$-93.0 = \Delta H^\circ_f\ (PCl_5) - (-306 + 0)$

13.39 $2\ Al(s) + 3\ O_2(g) \rightarrow Al_2O_3(s)$

At 25°C when 102 g of aluminum is burned at constant pressure in an atmosphere of oxygen to give Al_2O_3, 3150 kJ of heat energy is liberated. The standard enthalpy of formation of Al_2O_3 is

a. $-3150 \times \frac{102}{102}$ b. $-3150 \div \frac{102}{27}$ c. $-3150 \times 102/27$

d. $-3150 \div \frac{102}{27 \times 2}$ e. -3150×27

13.40 What is the heat of formation of NO(g) in kJ/mol given the following information?

$$2NO(g) \rightarrow N_2(g) + O_2(g) + 180.\ kJ$$

a. -180. kJ b. -90. kJ c. 90. kJ d. 180. kJ e. 360. kJ

13.41 What is the change in enthalpy when 1.00 mol (30.0 g) of nitrogen(II) oxide is oxidized to nitrogen dioxide?

$$2NO(g) + O_2(g) \rightarrow 2NO_2(g) + 114\ kJ$$

a. -114 kJ b. -114/2 kJ c. 114/2 kJ d. 114 kJ e. 114 x 2 kJ

Hess' Law

13.42 The combustion of one mole of graphite and one mole of diamond in excess oxygen is represented by the following equations:

$$C\ (diamond) + O_2(g) \rightarrow CO_2(g) \quad \Delta H° = -395\ kJ$$

$$C\ (graphite) + O_2(g) \rightarrow CO_2(g) \quad \Delta H° = -393\ kJ$$

From a consideration of the data above, when 1 mole of diamond is converted to 1 mole of graphite

a. 2 kJ of heat are evolved to the surroundings.
b. 2 kJ of heat are absorbed from the surroundings.
c. 788 kJ are absorbed from the surroundings.
d. 788 kJ are evolved to the surroundings.
e. diamond is the stable allotropic form.

13.39 (d) $\dfrac{\dfrac{-\ 3150\ \text{kJ}}{1\ \text{Al}_2\text{O}_3}}{\dfrac{1\ \text{Al}_2\text{O}_3}{2\ \text{Al}} \times 102(\text{g})\ \text{Al} \times \dfrac{1\ \text{mol Al}}{27\ \text{g Al}}} = \dfrac{\dfrac{-3150}{102}}{27 \times 2}$

13.40 (c) $N_2(g) + O_2(g) \rightarrow 2\ NO$; $\Delta H = +180$ kJ
For one mole of product we get 180 kJ/2.

13.41 (b) For this reaction $\Delta H = -114$ kJ, since heat is shown as a product in the reaction. The thermochemical equation is for two moles of product. The question asks for one mole of product. Therefore, $\Delta H_{rxn} = -114$ kJ/2.

68% 13.42 (a)

C (diamond) + $O_2(g) \rightarrow CO_2(g)$	$\Delta H° = -395$ kJ
$CO_2(g) \rightarrow$ C (graphite) + $O_2(g)$	$\Delta H° = +393$ kJ
c (diamond) $\rightarrow$ c (graphite)	$\Delta H°_{rxn} = -2$ kJ

13.43 From the ΔH_f data calculate the ΔH of reaction for

$$Fe_2O_3(s) + 3CO(g) \rightarrow 3CO_2(g) + 2Fe(s)$$

	ΔH_f(kcal) at 1 atm and 298°K
Fe(s)	---
$Fe_2O_3(s)$	-196.5
CO(g)	- 26.4
$CO_2(g)$	- 94.1

a. -399.6 b. -128.8 c. +128.8 d. 6.6 e. -6.6

13.44 Given the following data:

	ΔH_f (kcal/mole)
$\frac{3}{2}O_2 + 2B \rightarrow B_2O_3$	-302
$O_3 + 2B \rightarrow B_2O_3$	-336

the heat for the reaction oxygen → ozone at 298°C and 1 atm pressure in kcal is

a. -336 b. -302 c. -638 d. -34 e. +34

13.45 You are given the following data:

$$2\ CO_2(g) \rightarrow 2\ CO(g) + O_2(g) \qquad \Delta H° = +566\ kJ$$

$$3\ CO(g) + O_3(g) \rightarrow 3\ CO_2(g) \qquad \Delta H° = -992\ kJ$$

What is the $\Delta H°$ for the process $3/2\ O_2(g) \rightarrow O_3(g)$? All are at 25°C and 1 atmosphere.

a. -1558 kJ b. -425 kJ c. +143 kJ d. +426 kJ e. +1558 kJ

13.43 (e) $\Delta H_{rxn} = \Sigma\Delta H°_f\,(\text{prod.}) - \Sigma\Delta H°_f\,(\text{react.})$

$$= 3\,(-94.1) - (-196.5) - 3\,(-26.4)$$
$$= -6.6 \text{ kcal}$$

13.44 (e)

	ΔH_f(kcal/mol)
$3/2\ O_2 + 2B \rightarrow B_2O_3$	$-$ 302 kcal
$B_2O_3 \rightarrow O_3 + 2B$	+ 336 kcal
$3/2\ O_2 \rightarrow O_3$	+ 34 kcal

13.45 (c)

$3/2\ O_2(g) + 3\ CO(g) \rightarrow 3\ CO_2(g)$	3/2 (-566 kJ)
$3\ CO_2(g) \rightarrow 3\ CO(g) + O_3(g)$	+992 kJ
$3/2\ O_2(g) \rightarrow O_3(g)$	+143 kJ

13.46 Calculate the standard heat of formation of $CaCO_3(s)$ given

$$CaO(s) + CO_2(g) \rightarrow CaCO_3(s); \quad \Delta H = -178.1 \text{ kJ}$$

and

Compound	ΔH°_f, kJ mol^{-1}
$CO_2(g)$	-393.5
$CaO(s)$	-635.5

a. -1207 kJ b. -851 kJ c. -420 kJ d. +851 kJ e. +1207 kJ

13.47 Given the following at 25°C and 1.00 atm

	ΔH° kJ
$Pb(s) + PbO_2(s) + 2\ H_2SO_4(\ell) \rightarrow 2\ PbSO_4(s) + 2\ H_2O(\ell)$	-509.2
$SO_3(g) + H_2O(\ell) \rightarrow H_2SO_4(\ell)$	-130

Evaluate ΔH° for the reaction below at 25°C:

$$Pb(s) + PbO_2(s) + 2\ SO_3(g) \rightarrow 2\ PbSO_4(s)$$

a. -3.77×10^3 kJ b. $+3.77 \times 10^3$ kJ c. -639 kJ d. -521 kJ
e. -769 kJ

13.48 Calculate the standard heat of formation, ΔH°_f, of $CH_3OH(\ell)$, from the measured heat of combustion below and the other tabulated ΔH°_f values.

$$2CH_3OH(\ell) + 3O_2(g) \rightarrow 2CO_2(g) + 4H_2O(\ell); \quad \Delta H^\circ = -1454 \text{ kJ}$$

$\Delta H^\circ_f\ CO_2(g) = -394$ kJ mol^{-1}

$\Delta H^\circ_f\ H_2O(\ell) = -286$ kJ mol^{-1}

a. -478 kJ b. -239 kJ c. 239 kJ d. 478 kJ e. 1932 kJ

44% 13.46 (a) $\Delta H_{rxn} = \Delta H^\circ_f\ (CaCO_3) - (\Delta H^\circ_f\ (CaO) + \Delta H^\circ_f\ (CO_2))$

$-178.1\ kJ = \Delta H^\circ_f\ (CaCO_3) - (-635.5 - 393.5)$

$\Delta H^\circ_f\ (CaCO_3) = -1207\ kJ$

13.47 (e)

	ΔH° (kJ)
$Pb(s) + PbO_2(s) + 2H_2SO_4(\ell) \rightarrow 2PbSO_4(s) + 2H_2O(\ell)$	-509.2
$2SO_3(g) + 2H_2O(\ell) \rightarrow 2H_2SO_4(\ell)$	2(-130)
$Pb(s) + PbO_2(s) + 2SO_3(g) \rightarrow 2PbSO_4(s)$	-769 kJ

36% 13.48 (b) $\Delta H^\circ_{rxn} = 2\Delta H^\circ_f(CO_2) + 4\Delta H^\circ_f(H_2O) - 2\Delta H^\circ_f(CH_3OH)$

$-1454\ kJ = 2(-394) + 4(-286) - 2\Delta H^\circ_f(CH_3OH)$

$2\Delta H^\circ_f(CH_3OH) = -478\ kJ$

$\Delta H^\circ_f(CH_3OH) = -239\ kJ$

13.49 Given the following data:

$N_2O(g) + 3\ H_2(g) \rightarrow N_2H_4(\ell) + H_2O(\ell) \qquad \Delta H° = -317.0\ kJ$

$\Delta H_f°\ (N_2O(g)) = 82.1\ kJ$

$\Delta H_f°\ (H_2O(\ell)) = -285.8\ kJ$

the standard heat of formation of hydrazine, $\Delta H_f°(N_2H_4(\ell))$, is

a. -317.0 +82.1 +285.8

b. -317.0 +82.1 -285.8

c. -317.0 -82.1 +285.8

d. 317.0 -82.1 -285.8

e. 317.0 +82.1 -285.8

13.50 Using the following information

$Fe_2O_3(s) + 3\ CO(g) \rightarrow 2\ Fe(s) + 3\ CO_2(g) \qquad \Delta H = -26.8\ kJ$

$FeO(s) + CO(g) \rightarrow Fe(s) + CO_2(g) \qquad \Delta H = -16.5\ kJ$

the ΔH value for the reaction

$Fe_2O_3(s) + CO(g) \rightarrow 2\ FeO(s) + CO_2(g)$

is

a. -43.3 kJ b. -10.3 kJ c. 6.2 kJ d. 10.3 kJ e. 22.7 kJ

13.51 The compound $SO_2(g)$ can be oxidized to $SO_3(g)$ in the presence of excess oxygen. A total of 196 kJ of energy is liberated when 2.00 mole of SO_2 is oxidized. What is $\Delta H_f°$ of $SO_3(g)$? ($\Delta H_f°\ SO_2(g) = -297$ kJ/mol)

a. -790 kJ b. -395 kJ c. -98 kJ d. 199 kJ e. 398 kJ

13.52 The bond energies for $H_2(g)$ and $HCl(g)$ are 435 kJ/mole and 431 kJ/mole respectively. For the reaction $1/2\ H_2(g) + 1/2\ Cl_2(g) \rightarrow HCl(g)$, the heat of reaction is -92 kJ/mole of HCl produced. The bond energy of Cl_2 in kJ/mole is

a. 88 b. 96 c. 188 d. 243 e. 335

70% 13.49 (a)

	$\Delta H°(kJ)$
$N_2O(g) + 3H_2(g) \rightarrow N_2H_4(\ell) + H_2O(\ell)$	-317.0
$N_2(g) + 1/2\ O_2(g) \rightarrow N_2O(g)$	+ 82.1
$H_2O(g) + \quad \rightarrow H_2(g) + 1/2\ O_2(g)$	+285.8
$N_2(g) + 2H_2(g) \rightarrow N_2H_4(\ell)$	+ 50.9 kJ

or $\Delta H°_{rxn} = \Delta H°_f\ (N_2H_4) + \Delta H°_f(H_2O) - \Delta H°_f(N_2O)$

$-317.0\ kJ = \Delta H°_f(N_2H_4) - 285.8\ kJ - 82.1\ kJ$

80% 13.50 (c)

	$\Delta H(kJ)$
$Fe_2O_3 + 3CO \rightarrow 2Fe + 3CO_2$	-26.8
$2Fe + 2CO_2 \rightarrow 2FeO + 2CO$	2(+16.5)
$Fe_2O_3 + CO \rightarrow 2FeO + CO_2$	6.2 kJ

52% 13.51 (b) $2SO_2(g) + O_2(g) \rightarrow 2SO_3$, $\Delta H = -196$ kJ

$\Delta H_{rxn} = 2\Delta H°_f(SO_3) - 2\Delta H°_f(SO_2)$

$-196\ kJ = 2\Delta H°_f(SO_3) - 2(-297\ kJ)$

$2\Delta H°_f(SO_3) = -790\ kJ$

$\Delta H°_f(SO_3) = -395\ kJ$

13.52 (d) $\Delta H_{rxn} = 1/2\ B.E.\ (H_2) + 1/2\ B.E.\ (Cl_2) - B.E.\ (HCl)$

$-92 = \frac{435}{2} + 1/2\ B.E.\ (Cl_2) - 431$

$1/2\ B.E.\ (Cl_2) = 121.5\ kJ$

$B.E.\ (Cl_2) = 243\ kJ$

Entropy

13.53 The total entropy of a system and its surroundings always increases for a spontaneous process. This represents

a. the law of conservation of energy.

b. the law of conservation of matter.

c. in words the first law of thermodynamics.

d. in words the second law of thermodynamics.

e. in words the third law of thermodynamics.

13.54 Arrange the following in order of <u>increasing</u> entropy:

$Hg(\ell)$, $Hg(s)$, $C_6H_6(\ell)$, $CH_3OH(\ell)$

a. $Hg(s)$, $CH_3OH(\ell)$, $C_6H_6(\ell)$, $Hg(\ell)$

b. $CH_3OH(\ell)$, $Hg(s)$, $Hg(\ell)$, $C_6H_6(\ell)$

c. $Hg(\ell)$, $Hg(s)$, $C_6H_6(\ell)$, $CH_3OH(\ell)$

d. $Hg(s)$, $Hg(\ell)$, $C_6H_6(\ell)$, $CH_3OH(\ell)$

e. $Hg(s)$, $Hg(\ell)$, $CH_3OH(\ell)$, $C_6H_6(\ell)$

13.55 Which of the following would be expected to have the largest entropy value per mole?

a. $SO_2Cl_2(s)$ b. $SO_2Cl_2(\ell)$ c. $SO_2Cl_2(g)$ d. $SO_2(g)$

e. $Cl_2(g)$

13.56 Arrange the following in order of <u>increasing</u> entropy:

$CH_4(g)$, $C(s)$, $Li(s)$, $Na(s)$

a. Li, C, CH_4, Na b. C, Li, Na, CH_4 c. Na, Li, C, CH_4

d. CH_4, Li, Na, C e. Na, Li, CH_4, C

13.53 (d) This represents one version of the second law of thermodynamics. There are many statements of the second law and a few are the following: "Heat cannot spontaneously pass from a colder to a warmer body." "Every system which is left to itself will, on the average, change toward a condition of maximum probability." "Gain in information is loss in entropy."

13.54 (e) Consider only (d) and (e) because the entropy of Hg(ℓ) must be greater than Hg(s). C_6H_6 is a bigger molecule (has more atoms) than CH_3OH and there is intermolecular hydrogen bonding in CH_3OH which will give it more order.

13.55 (c) Gases have larger entropy values than solids or liquids. The larger the gas molecule the greater is its entropy value.

13.56 (b) Since CH_4 is a gas, the responses that we need to consider are (b) and (c). Both Li and Na are in the same family and Na is larger and has a lower melting point than Li so it will have a larger entropy value.

13.57 For which of the following reactions is $\Delta S° < 0$?

a. $MgCO_3(s) \rightarrow MgO(s) + CO_2(g)$

b. $NH_4Br(s) \rightarrow NH_3(g) + HBr(g)$

c. $CH_3OH(\ell) \rightarrow CH_3OH(g)$

d. $2NO_2(g) \rightarrow 2NO(g) + O_2(g)$

e. $2BrCl(g) \rightarrow Cl_2(g) + Br_2(\ell)$

13.58 In which one of the following processes would one expect ΔS to have a value closest to zero?

a. $H_2(g) + F_2(g) \rightarrow 2HF(g)$

b. $C_2H_4(g) + HBr(g) \rightarrow C_2H_5Br(g)$

c. $CO(g) + 1/2\ O_2(g) \rightarrow CO_2(g)$

d. $2NO(g) + O_2(g) \rightarrow 2\ NO_2(g)$

e. $CH_3C(O)H(g) + 5/2\ O_2(g) \rightarrow 2\ CO_2(g) + 2H_2O(g)$

13.59 Which of the following processes would be expected to have the smallest ΔS value?

a. $CO_2(s) \rightarrow CO_2(g)$ b. $CO_2(s) \rightarrow CO_2(\ell)$

c. $2C(s) + O_2(g) \rightarrow 2CO(g)$ d. $C(s) + CO_2(g) \rightarrow 2CO(g)$

e. $C(s) + 1/2\ O_2(g) \rightarrow CO(g)$

13.60 Which one of the following processes would be expected to have a ΔS value very close to zero?

a. $H_2O(s) \rightarrow H_2O(\ell)$ b. $2H_2(g) + O_2(g) \rightarrow 2H_2O(g)$

c. $H_2O(s) \rightarrow H_2O(g)$ d. $N_2(g) + O_2(g) \rightarrow 2NO(g)$

e. $OF_2(g) + H_2O(g) \rightarrow O_2(g) + 2HF(g)$

85% 13.57 (e) Reactions in which the number of moles of gaseous molecules increases usually have positive values of $\Delta S°$. Δn_{gas} is positive for reactions (a) - (d). $\Delta n_{gas} = 1 - 2 = -1$ for reaction (e).

13.58 (a) $\Delta n_{gas} = 0$

(b) $\Delta n = -1$; ΔS will be negative.

(c) $\Delta n = -1/2$; ΔS will be negative.

(d) $\Delta n = -1$; ΔS will be negative.

(e) $\Delta n = +1/2$; ΔS will be positive.

13.59 (b) Which process has a $\Delta n_{gas} = 0$?

13.60 (d) Look for a process which has the same number of moles of gaseous products as there are moles of gaseous reactants.

13.61 The heat of fusion of acetic acid (M.W. = 60.0) is 192 J/g. Its melting point is 16.6°C. The change in entropy for the melting of acetic acid in J/mol•K is

a. -11.6 b. 11.6 c. 15.1 d. 39.8 e. 694

13.62 The dissolution of potassium chloride crystals in water is depicted by the equation

$$KCl(s) \rightarrow K^{+}(aq) + Cl^{-}(aq) \qquad \Delta H = 17 \text{ kJ mol}^{-1}$$

The driving force in the dissolution process is because

a. the KCl crystals are unstable.

b. all processes move spontaneously from low to higher energy states.

c. all processes move spontaneously from high to lower energy states.

d. the system becomes more disordered.

e. the energy of the hydrated ions is lower than the energy of the ions in the crystal lattice.

Free Energy and Reaction Spontaneity

13.63 All of the following have free energy of formation values of zero EXCEPT

a. Li(s) b. He(g) c. $N_2(g)$ d. $F_2(g)$ e. $I_2(g)$

13.64 If a process is both endothermic and spontaneous then

a. $\Delta E<0$ b. $\Delta H<0$ c. $\Delta G>0$ d. $\Delta S>0$ e. $\Delta H=0$

13.65 Which of the following is true for the reaction $NH_3(\ell) \rightleftarrows NH_3(g)$ at -33°C and 1-atm pressure (the normal boiling point for NH_3 is -33°C)?

a. $\Delta H = 0$ b. $\Delta S = 0$ c. $\Delta H = P\Delta V$ d. $\Delta H = \Delta nRT$ e. $\Delta H = T\Delta S$

70% 13.61 (d) $\Delta S = \frac{\Delta H_{fusion}}{T_{(freezing\ point)}} = \frac{192\ J/g \times 60.0\ g/mol}{289.6\ K}$

52% 13.62 (d) The change in entropy is the driving force. 32% of the students selected response (e). This response is incorrect because energy of the hydrated ions is higher because the process is endothermic.

13.63 (e) At 25°C and one atmosphere pressure, I_2 is a solid and has a $\Delta G°_f$ value of zero.

60% 13.64 (d) A process is spontaneous when ΔG is negative. From a consideration of the equation, ΔG = ΔH - TΔS, ΔS must be positive when ΔH is positive in order to have a negative free energy value.

80% 13.65 (e) At the boiling point of a liquid and at 1-atm pressure, a reaction system is in equilibrium and ΔG is zero. Therefore, 0 = ΔH - TΔS.

13.66 The combination of ΔH and ΔS values which will <u>always</u> give ΔG a negative value is

	ΔH	ΔS
a.	+	+
b.	+	-
c.	-	+
d.	-	-
e.	+ or -	+

13.67 The combination of ΔH and ΔS values which <u>will always</u> favor the forward direction of a given reaction is

	ΔH	ΔS
a.	+	+
b.	-	-
c.	-	+
d.	+	-
e.	+	+ or -

13.68 The combination of ΔH and ΔS values which <u>will always</u> favor the <u>reverse</u> direction of a given reaction is

	ΔH	ΔS
a.	+	+
b.	-	-
c.	-	+
d.	+	-
e.	-	+ or -

13.69 For the following reaction which is spontaneous at 1600°C

$$Br_2(g) \rightarrow 2Br(g)$$

a. ΔH is + and ΔS is + for the reaction.

b. ΔH is - and ΔS is - for the reaction.

c. ΔH is + and ΔS is - for the reaction.

d. ΔH is - and ΔS is + for the reaction.

e. ΔG is - for all temperatures.

80% 13.66 (c) $\Delta G = \Delta H - T\Delta S$

$- = (-) - (+)$

13.67 (c) The forward reaction is always favored when ΔG is negative. ΔG will always be negative when ΔH is negative and ΔS is positive. There is always a drive to go to a minimum energy and a maximum randomness. $\Delta G = \Delta H - T\Delta S$

64% 13.68 (d) The reverse process will occur when ΔG is positive.

$\Delta G = \Delta H - T\Delta S$

ΔG will always be positive when ΔH is positive and ΔS is negative.

$+ = + - (-)$

13.69 (a) Bond breaking is an endothermic process so ΔH is +. Since $\Delta n_{gas} = 2 - 1 = +1$, then ΔS is positive for the process.

13.70 Sublimation is an example of a process for which

a. ΔH, ΔS, and ΔG are negative at all temperatures.
b. ΔH, ΔS, and ΔG are positive at all temperatures.
c. ΔH and ΔS are positive at all temperatures.
d. ΔH is negative and ΔS is negative at all temperatures.
e. ΔH is negative and ΔS is positive at all temperatures.

13.71 Condensation is an example of a process for which

a. ΔH, ΔS, and ΔG are negative at all temperatures.
b. ΔH, ΔS, and ΔG are positive at all temperatures.
c. ΔH and ΔS are positive.
d. ΔH is negative and ΔS is positive.
e. ΔH is negative and ΔS is negative.

13.72 For the reaction, $2NO(g) + O_2(g) \rightarrow 2NO_2(g)$ at 1 atmosphere pressure, the values of ΔH and ΔS are both negative and the process is spontaneous at room temperature. Which of the following is also true?

a. The direction of the reaction may be reversed at high temperature.
b. The change in entropy is the driving force.
c. At high temperatures ΔS becomes positive.
d. At high temperatures ΔH becomes positive.
e. ΔG is temperature independent.

13.73 For the reaction, $H_2(g) + S(s) \rightarrow H_2S(g)$ at 1 atmosphere pressure, the value of ΔH is negative and ΔS is positive and the reaction is spontaneous at room temperature. Which of the following statements concerning this reaction is true?

a. ΔG is positive at room temperature.
b. The reverse reaction will go at high temperatures.
c. The reverse reaction will go at low temperatures.
d. The sign of ΔG will change with temperature.
e. This reaction will only go to the right.

64% 13.70 (c) ΔH_{fusion} , $\Delta H_{vaporization}$ and $\Delta H_{sublimation}$ are all endothermic processes. Δn_{gas} is positive for the process, $M(s) \rightarrow M(g)$, so ΔS is positive.

64% 13.71 (e) If vaporization is endothermic then condensation must be exothermic. Δn_{gas} is negative so ΔS is negative.

56% 13.72 (a) From a consideration of the equation, $\Delta G = \Delta H - T\Delta S$, at high temperatures the $T\Delta S$ term can become larger than ΔH. If ΔS is negative and $T\Delta S$ becomes larger than ΔH (and ΔH is negative) ΔG will become positive.

84% 13.73 (e) When the signs of ΔH and ΔS are opposite to each other, a temperature change cannot affect the direction of the reaction.

13.74 For the following reaction which is spontaneous at all temperatures

$$CaC_2(s) + 2\ H_2O(\ell) \rightarrow Ca(OH)_2(s) + C_2H_2(g)$$

one would predict that

a. ΔG is + for all temperatures.

b. ΔH is + and ΔS is + for the reaction.

c. ΔH is - and ΔS is - for the reaction.

d. ΔH is - and ΔS is + for the reaction.

e. ΔH is + and ΔS is - for the reaction.

13.75 At 2200 K, the reaction $CaO(s) + SO_3(g) \rightarrow CaSO_4(s)$ is nonspontaneous whereas it is spontaneous at room temperature. All of the following statements concerning this reaction are true EXCEPT

a. ΔG is negative at room temperature.

b. Both ΔH and ΔS are negative for the reaction.

c. ΔG becomes zero at a temperature between 300 K and 2200 K.

d. The change in enthalpy is the main driving force of the reaction.

e. The change in entropy is the main driving force of the reaction.

13.76 If the entropy change for a chemical reaction at 27°C is -10 J/mol·K and ΔH for the reaction is -20 kJ, ΔG is

a. +2980 J b. -3020 J c. -17,000 J d. -19,730 J

e. -23,000 J

13.77 At 25°C for the reaction $2\ SO_2(g) + O_2(g) \rightleftarrows 2\ SO_3(g)$ $\Delta H°$ is -197.7 kJ and $\Delta S°$ is -188 J K^{-1}. What is ΔG for this reaction?

a. -253.7 kJ b. -193.0 kJ c. -141.7 kJ d. 4.502×10^3 kJ

e. 5.583×10^4 kJ

13.74 (d) Since the reaction is spontaneous at all temperatures, ΔG must be negative at all temperatures.

$\Delta G = \Delta H - T\Delta S$

$- = ? -(?)$

ΔS would be expected to be positive since $\Delta n_{gas} = +1$.

$\therefore - = ? -(+)$

At very low temperatures, the term $T\Delta S$ becomes zero, therefore, for ΔG to be negative, ΔH must be negative.

13.75 (e) At room temperature, ΔG is negative. ΔS is negative for the process because $\Delta n_{gas} = -1$.

$\Delta G = \Delta H - T\Delta S$

$- = ? -(-)$

So that ΔG is negative, ΔH must be negative.

13.76 (c) $\Delta G = \Delta H - T\Delta S$

$= -20{,}000\ J/mol - 300\ K\ (10\ J/mol\cdot K)$

$= -23{,}000\ J$

Before doing any calculations, make sure that there is compatibility of the ΔH and ΔS units.

65% 13.77 (c) $\Delta G = \Delta H - T\Delta S$

$= -197{,}700\ J - 298\ K\ (-188\ J/K)$

13.78 For the reaction

$$3C(s) + 4H_2(g) \rightleftarrows C_3H_8(g)$$

$\Delta S° = -0.269$ kJ/mol•K

$\Delta H° = -103.8$ kJ/ mol

Calculate the equilibrium constant at 25°C for the reaction above.

a. 1.0 b. 1.4×10^4 c. 1.0×10^{17} d. 3.7×10^{19}

e. 2.1×10^{32}

40% 13.78 (b) $\Delta G° = \Delta H° - T\Delta S°$

$= -103.8\ kJ/mol - 298\ K\ (-0.269\ kJ/mol\cdot K)$

$= -23.64\ kJ/mol$

$K = 10^{-\Delta G°/2.3\ RT}$

$K = 10^{23640/2.303 \times 8.314 \times 298}$

$K = 1.4 \times 10^{4}$

OXIDATION-REDUCTION AND ELECTROCHEMISTRY

Oxidation and Reduction

14.1 Which of the following changes requires a reducing agent?

a. $Cr_2O_7^{2-} \rightarrow CrO_4^{2-}$ b. $ClO_4^- \rightarrow ClO_3^-$ c. $NH_3 \rightarrow NH_4^+$
d. $Al \rightarrow Al^{3+}$ e. $O^{2-} \rightarrow O_2^{2-}$

14.2 Which of the following changes requires an oxidizing agent?

a. $NH_3 \rightarrow NH_4^+$ b. $N_2H_4 \rightarrow N_2$ c. $NO_3^- \rightarrow N_2O$ d. $NO_2^- \rightarrow N_2O_3$
e. $NO_2 \rightarrow N_2O_4$

14.3 All the following are considered as oxidizing agents EXCEPT

a. H_2O_2 b. concentrated HNO_3 c. Cl_2 d. MnO_2 e. CO_2

14.4 Which species would be expected to function as a reducing agent?

a. NO_3^- b. Mg^{2+} c. Ti^{3+} d. Na^+ e. ClO_4^-

14.5 All of the following species would be expected to function as both oxidizing agents and reducing agents EXCEPT:

a. ClO^- b. Cl_2 c. V^{3+} d. Cl^- e. NO_2

14.6 In the following oxidation-reduction reaction

$$8H^+(aq) + 6Cl^-(aq) + Sn(s) + 4NO_3^-(aq) \rightarrow SnCl_6^{2-}(aq) + 4NO_2(g) + 4H_2O$$

the oxidizing agent is

a. H^+ b. Cl^- c. Sn   d. NO_3^- e. $SnCl_6^{2-}$

60% 14.1 (b) A reducing agent decreases the oxidation number of an atom in a compound or ion. A reducing agent provides electrons to the substance that is reduced.

$$\underset{+7}{ClO_4^-} + 2e \rightarrow \underset{+5}{ClO_3^-}$$

50% 14.2 (b) An oxidizing agent increases an oxidation number of an atom in a compound. An oxidizing agent removes electrons from the substance that is oxidized.

$$\underset{2(-2)}{N_2H_4} \rightarrow \underset{2(0)}{N_2} + 4e$$

30% 14.3 (e) An oxidizing agent has a large attraction for electrons. The central atom in many oxidizing agents has a high oxidation number. H_2O_2, HNO_3, Cl_2 and MnO_2 are all well-known oxidizing agents.

14.4 (c) Look for a species which is not in its highest oxidation state. A species in its highest oxidation state does not have electrons available for donation. The highest oxidation state of Ti is 4+.

14.5 (d) Species that can function as both oxidizing and reducing agents are in intermediate oxidation states. Cl^- can only function as a reducing agent since it is in its lowest oxidation state.

14.6 (d) The oxidation number of the oxidizing agent is decreased.

$$\underset{+5}{NO_3^-} + e \rightarrow \underset{+4}{NO_2}$$

14.7 In the balanced equation

$$2\ MnO_4^- + 5\ C_2O_4^{2-} + 16\ H^+ \rightarrow 2\ Mn^{2+} + 10\ CO_2 + 8\ H_2O$$

the reducing agent is

a. MnO_4^- b. $C_2O_4^{2-}$ c. H^+ d. Mn^{2+} e. CO_2

14.8 In which of the following half reactions is oxidation occurring?

a. $N_2O_5 \rightarrow NO_3^-$ b. $CrCl_2 \rightarrow CrCl_3$ c. $P_4O_{10} \rightarrow H_3PO_4$
d. $MnO_4^- \rightarrow MnO_4^{2-}$ e. $H_2O \rightarrow H_2$

14.9 A sulfur containing species that cannot be reduced is

a. H_2SO_3 b. SO_4^{2-} c. $S_2O_3^{2-}$ d. SO_3^{2-} e. S^{2-}

14.10 In which of the following species does the underlined element have an oxidation number of +2?

a. $\underline{Zn}(OH)_4^{2-}$ b. $\underline{Cr}O_2Cl_2$ c. $H\underline{N}O_2$ d. $\underline{P}H_4^+$ e. $\underline{N}O_2$

14.11 In which of the following half reactions is reduction occurring?

a. $SO_3 \rightarrow SO_4^{2-}$ b. $SO_2 \rightarrow HSO_3^-$ c. $H_2O_2 \rightarrow OH^-$

d. $Mn^{2+} \rightarrow Mn_2O_3$ e. $CrO_4^{2-} \rightarrow Cr_2O_7^{2-}$

14.12 All of the following half-reactions are balanced EXCEPT

a. $2H_2O + 2e \rightarrow H_2 + 2OH^-$
b. $NO_3^- + 4H^+ + 3e \rightarrow NO + 2H_2O$
c. $2Ta + 5H_2O \rightarrow Ta_2O_5 + 10H^+ + 10e$
d. $H_3PO_3 + H_2O \rightarrow H_3PO_4 + 2H^+ + 2e$
e. $Fe^{3+} \rightarrow Fe^{2+} + e$

14.7 (b) The reducing agent is oxidized.

$$\underset{+3}{C_2O_4^{2-}} \rightarrow \underset{+4}{2CO_2} + 2e$$

76% 14.8 (b) Cr changes from 2+ to 3+. Oxidation is an increase in oxidation number.

80% 14.9 (e) The sulfur in the sulfide ion is in its lowest oxidation so it can not accept any more electrons as it already has a noble gas configuration.

92% 14.10 (a) Let y represent the oxidation number of Zn in $Zn(OH)_4^{2-}$.

$$y + 4\ (-2) + 4\ (+1) = -2$$
$$y = +2$$

or

$$y + 4\ OH^- = -2$$
$$y + 4\ (-1) = -2\ ;\ y = +2$$

68% 14.11 (c) $\underset{-1}{H_2O_2} + 2e \rightarrow \underset{-2}{2OH^-}$

The oxidation number of oxygen changes from -1 to -2. This is reduction.

14.12 (e) The half-reaction in (e) does not have a charge balance.

$$+3 \neq +2 - 1$$

14.13 $I_2 + 6H_2O \rightarrow 2IO_3^- + 12H^+ +$ _____ e

How many electrons are needed to balance the half-reaction above?

a. 2 b. 6 c. 8 d. 10 e. 12

14.14 For the oxidation of ethanol to ethanoic acid (acetic acid)

$$C_2H_5OH \rightarrow CH_3COOH \quad \text{(not balanced)}$$

the number of electrons that must be added to the right side is

a. zero b. one c. two d. three e. four

Questions 14.15 and 14.16 - balance each equation. Choose from the numbers given as answers the one which is the coefficient of the underlined substance in the balanced ionic reaction.

14.15 $Zn + NO_3^- + H^+ \rightarrow Zn^{2+} + NH_4^+ + \underline{H_2O}$

a. 1 b. 2 c. 3 d. 4 e. 5

14.16 $MnO_4^- + H^+ + Cl^- \rightarrow Mn^{2+} + \underline{Cl_2} + H_2O$

a. 1 b. 2 c. 3 d. 4 e. 5

14.17 The following reaction is used for the analytical determination of uranium in <u>acidic</u> solution:

$$UO^{2+} + Cr_2O_7^{2-} \rightarrow UO_2^{2+} + Cr^{3+} \quad \text{(unbalanced)}$$

For every mole of UO^{2+} in an unknown _____ mole(s) of $Cr_2O_7^{2-}$ would be required for the above reaction.

a. 1/3 b. 1/2 c. 2 d. 3 e. 6

14.18 Balance the equation below. Choose from the numbers given as answers the one which is the coefficient of the underlined substance in the balanced ionic reaction:

$$\underline{Zn} + NO_3^- \rightarrow Zn(OH)_4^{2-} + NH_3 \quad \text{(basic solution)}$$

a. 1 b. 2 c. 4 d. 6 e. 8

14.13 (d) To balance this half-reaction, we need only to balance the charges. The left hand side of equation has zero charge, therefore, we get 0 = -2 + 12 + y ; y = -10 or 10 electrons.

60% 14.14 (e) $C_2H_5OH \rightarrow CH_3COOH$

First, balance the oxygen atoms by adding one water to the left hand side.

$$C_2H_5OH + H_2O \rightarrow CH_3COOH$$

Now balance the hydrogen atoms by adding H^+ to the right hand side.

$$C_2H_5OH + H_2O \rightarrow CH_3COOH + 4\,H^+$$

Now balance the charge by adding electrons to the appropriate side.

$$C_2H_5OH + H_2O \rightarrow CH_3COOH + 4\,H^+ + 4\,e$$

14.15 (c)

$$4(Zn \rightarrow Zn^{2+} + 2e)$$
$$1\,(8e + NO_3^- + 10\,H^+ \rightarrow NH_4^+ + 3\,H_2O)$$

14.16 (e)

$$2(MnO_4^- + 5e + 8H^+ \rightarrow Mn^{2+} + 4H_2O)$$
$$5(2Cl^- \rightarrow Cl_2 + 2e\quad)$$

14.17 (a)

$$3(UO^{2+} \rightarrow UO_2^{2+} + 2e)$$
$$1(Cr_2O_7^{2-} + 6e \rightarrow 2\,Cr^{3+})$$

$\therefore$ 3 UO^{2+} / 1 $Cr_2O_7^{2-}$

$$\#\ \text{mole } Cr_2O_7^{2-} = \frac{1\ Cr_2O_7^{2-}}{3\ UO^{2+}} \times 1 \text{ mol } UO^{2+} = 1/3$$

14.18 (c)

$$4\,(Zn + 4\,OH^- \rightarrow Zn(OH)_4^{2-} + 2e$$
$$1(6H_2O + NO_3^- + 8e \rightarrow NH_3 + 9\,OH^-)$$

The information below is for questions 14.19 and 14.20.

$CrO_4^{2-} + HSnO_2^- \rightarrow CrO_2^- + HSnO_3^-$ (basic solution, unbalanced equation)

14.19 The oxidation numbers of chromium and tin in the reactants and products are respectively:

a. 6, 4, 3, 6 b. 6, 2, 3, 4 c. 6, 2, 4, 4 d. 2, 1, 1, 1
e. 4, 2, 2, 3

14.20 In the balanced equation for the reaction above, the coefficients for CrO_4^{2-} and $HSnO_2^-$ are, respectively:

a. 1,3 b. 1,2 c. 2,1 d. 3,2 e. 2,3

14.21 The following reaction occurs in basic solution:

$\_\_H_2O + \_\_MnO_4^- + \_\_ClO_2^- \rightarrow \_\_MnO_2 + ClO_4^- + \_\_OH^-$

Balance the equation above and then add together all of the coefficients in the balanced equation. The sum of these coefficients is:

a. 15 b. 20 c. 24 d. 25 e. 30

14.22 For the reaction between permanganate ion and sulfite ion in basic solution, the unbalanced equation is

$$MnO_4^- + SO_3^{2-} \rightarrow MnO_2 + SO_4^{2-}.$$

When this equation is balanced using the smallest whole number coefficients possible, the number of OH^- ions is

a. two on the right. b. two on the left. c. three on the right.
d. four on thc right. e. four on the left.

14.19 (b) CrO_4^{2-} ; $y + 4(-2) = -2$; $y = +6$

$HSnO_2^-$; $y + 1 + 2(-2) = -1$; $y = +2$

CrO_2^- ; $y + 2(-2) = -1$; $y = +3$

$HSnO_3^-$; $y + 1 + 3(-2) = -1$; $y = +4$

14.20 (e) $2(CrO_4^{2-} + 3e \rightarrow CrO_2^-)$

$3(HSnO_2^- \rightarrow HSnO_3^- + 2e)$

14.21 (b) $4(2H_2O + MnO_4^- + 3e \rightarrow MnO_2 + 4OH^-)$

$\underline{3(ClO_2^- + 4OH^- \rightarrow ClO_4^- + 4e + 2H_2O}$

$$8H_2O + 4MnO_4^- + 3ClO_2^- + 12OH^- \rightarrow 4MnO_2 + 3ClO_4^- + 16OH^- + 6H_2O$$

$$2H_2O + 4MnO_4^- + 3ClO_2^- \rightarrow 4MnO_2 + 3ClO_4^- + 4OH^-$$

60% 14.22 (a) $3(SO_3^{2-} + 2OH^- \rightarrow SO_4^{2-} + H_2O + 2e)$

$\underline{2(2H_2O + MnO_4^- + 3e \rightarrow MnO_2 + 4OH^-)}$

Galvanic Cells and Electrolytic Cells

14.23 In a galvanic cell, the electron flow is always from

a. the positive electrode to the negative electrode.

b. through the salt bridge.

c. from the oxidizing agent through the salt bridge to the reducing agent.

d. from the reducing agent through the salt bridge to the oxidizing agent.

e. from the anode to the cathode.

14.24 In an electrolytic cell, the electrode that acts as a source of electrons <u>to the solution</u> is called the ________________. The chemical change that occurs at this electrode is called ________________.

a. cathode, reduction b. cathode, oxidation c. anode, oxidation

d. anode, reduction e. Cannot tell unless we know the species that are involved in the cell reaction.

14.25 The numerical value of the Faraday constant is given as 96,500. This value represents

a. the number of coulombs carried by one mole of electrons.

b. the number of electrons corresponding to one coulomb of charge.

c. the number of electrons corresponding to one mole of electric charge.

d. the number of ions discharged by the passage of one mole of electrons.

e. $96{,}500\ \frac{\text{amp}}{\text{sec}}$

14.26 Which of the following processes could occur at the anode of an electrochemical cell?

a. $Cu^{2+} + 2e \rightarrow Cu$ b. $Zn^{2+} + 2e \rightarrow Zn$ c. $Cu \rightarrow Cu^{2+} + 2e$

d. $Zn \rightarrow Zn^{2+} + 2e$ e. both c. and d.

14.23 (e) Oxidation occurs at the anode and reduction at the cathode so electrons must flow from the anode to the cathode.

42% 14.24 (a) The cathode provides electrons to the species that is undergoing reduction.

64% 14.25 (a) One mole of electrons equals one Faraday = 96,500 coulombs = 96,500 amp. sec.

14.26 (e) Oxidation occurs at the anode. Oxidation is an increase in oxidation number. Oxidation is a loss of electrons.

$$Cu \rightarrow Cu^{2+} + 2e$$

$$Zn \rightarrow Zn^{2+} + 2e$$

We can see from the two equations above that there is both a loss of electrons and an increase in oxidation numbers.

14.27 Molten magnesium chloride is electrolyzed using inert electrodes and reactions represented by the following equations occur:

$$2\,Cl^- \rightarrow Cl_2 + 2e$$
$$Mg^{2+} + 2e \rightarrow Mg$$

Concerning this electrolysis, which of the following statements is TRUE?

a. Oxidation occurs at the cathode.

b. Mg^{2+} ions are reduced at the anode.

c. Electrons pass through the metallic part of the circuit from Mg^{2+} ions to the Cl^- ions.

d. Cl^- ions are oxidizing agents.

e. The cations in the electrolyte undergo reduction.

14.28 A steady current of 10.0 amp flowing for 965 sec. corresponds to

a. 9650 electrons b. 0.100 Faraday c. 0.100 volt

d. 0.100 coulomb e. None of these.

14.29 A current of 10 amperes is passed through molten magnesium chloride for 3.0 hours. How many moles of magnesium metal could be produced by this electrolysis (one Faraday = 96,500 coulombs)?

a. 1.1 b. 2.2 c. 0.56 d. 0.37 e. 0.22

14.30 How many Faradays of electricity are required to convert a mole of NO_3^- to NH_4^+?

a. 1 b. 3 c. 4 d. 5 e. 8

14.31 A copper electrode in $CuSO_4$ solution is connected by a wire to a silver electrode in a $AgNO_3$ solution. The two solutions are also joined by a salt bridge. Current is allowed to flow through the cell until the mass of the silver electrode has changed by 108 g. During this time the mass of the copper electrode will (atomic weights: Ag = 107.9; Cu 63.5)

a. decrease by 108 g b. increase by 108 g c. increase by 31.8 g

d. decrease by 31.8 g e. decrease by 63.5 g

76% 14.27 (e) (a) Reduction occurs at the cathode.

(b) Magnesium is oxidized to Mg^{2+} at the anode.

(c) Electrons pass from the chloride ions to the Mg^{2+} ions through the external circuit.

(d) Cl^- ions are reducing agents.

14.28 (b) $\# \mathcal{F} = 10.0 \text{ amp} \times 965 \text{ s} \times \frac{1 \mathcal{F}}{96{,}500 \text{ amp·s (coulomb)}}$

$\# \mathcal{F} = 0.100$ Faraday

14.29 (c) $\# \text{ mol e} = 10 \text{ amp} \times 3600 \frac{\text{s}}{\text{hr}} \times 3.0 \text{ hr} \times \frac{1 \text{ mol e}}{96{,}500 \text{ amp·s}}$

$= 1.12 \text{ mol e}$

$Mg^{2+} + 2e \rightarrow Mg$

$\# \text{ mol Mg} = \frac{1 \text{ Mg}}{2 \text{ e}} \times 1.12 \text{ mol e} = 0.56 \text{ mol}$

40% 14.30 (e) $10 H^+ + NO_3^- + 8e \rightarrow NH_4^+ + 3 H_2O$

From the balanced half-reaction above we can see that eight moles of electrons are required to reduce one mole of nitrate ions. The oxidation number change is from +5 to -3 for nitrogen.

35% 14.31 (d) $Cu + 2 Ag^+ \rightarrow Cu^{2+} + 2 Ag$

$\# \text{ mol Cu} = \frac{1 \text{ Cu}}{2 \text{ Ag}} \times \# \text{ mol Ag} = \frac{1}{2} \times 1 = 1/2$

$\# \text{ g Cu} = 0.5 \text{ mol Cu} \times \frac{63.5 \text{ g Cu}}{1 \text{ mol Cu}} = 31.8 \text{ g}$

More students selected answer (e) than those who selected (d). The first step in solving many chemistry problems is to write a balanced equation.

14.32 All the following statements concerning the electrolysis of water are TRUE EXCEPT

a. The volume of hydrogen produced is twice the volume of oxygen produced.

b. A direct current must be used in the electrolysis.

c. Hydrogen is evolved at anode.

d. Electrical energy must be supplied continuously.

e. An electrolyte such as some NaCl or H_2SO_4 must be added to the water.

14.33 When a dilute aqueous solution of Li_2SO_4 is electrolyzed, the products formed at the anode and cathode, respectively, are

a. O_2 and H_2 b. H_2 and O_2 c. O_2 and H^+ d. H_2 and OH^-

e. O_2 and Li

14.34 What is the cathode reaction during the electrolysis of aqueous NaCl?

a. $2Cl^- \rightarrow Cl_2 + 2e$ b. $2H_2O \rightarrow O_2 + 4H^+ + 4e$

c. $2H_2O + 2e \rightarrow H_2 + 2OH^-$ d. $Na^+ + e \rightarrow Na$

e. $4OH^- \rightarrow O_2 + 2H_2O + 4e$

14.35 One of the major products in the electrolysis of a concentrated aqueous sodium chloride solution is

a. Na b. O_2 c. NaOH d. Br_2 e. Mg

14.36 In purifying copper by electrolysis, the anode and cathode are made of ________________________________ respectively.

a. Pure copper and impure copper.

b. Pure copper and pure copper.

c. Impure copper and impure copper.

d. Impure copper and pure copper.

e. Only the anode needs to be pure copper.

14.32 (c) Hydrogen is obtained by reduction of water so hydrogen must be evolved from the cathode not the anode.

44% 14.33 (a) The Li_2SO_4 is an electrolyte that was added to help transport the current in order to facilitate the electrolysis of water.

$$2\ H_2O \xrightarrow{\text{elect.}} 2\ H_2 + O_2$$

Oxidation occurs at the anode (oxygen) and reduction occurs at the cathode (hydrogen). 40% of the students selected answer (b).

40% 14.34 (c) Water is reduced and hydrogen is evolved at the cathode.

45% 14.35 (c) The products of the electrolysis of molten NaCl are sodium and chlorine. The products of electrolysis of concentrated NaCl(aq) are chlorine, hydrogen and sodium hydroxide.

14.36 (d) The anode is impure copper which is put into solution and then deposited onto a pure copper cathode.

14.37 Which one of the following elements is obtained commercially by electrolytic reduction?

a. Chlorine b. Bromine c. Oxygen d. Magnesium e. Phosphorus

Electrode Potentials

14.38 A cell with a potential of 0.15 V has the cell reaction

$$2\ Cu^{2+}(aq) + H_2 \rightarrow 2\ Cu^{+}(aq) + 2\ H^{+}(aq)$$

If the concentrations of the ions were 1.0 molar and the pressure of H_2 were 1.0 atmosphere, then E° for the half reaction

$$Cu^{2+}(aq) + e \rightarrow Cu^{+}(aq)$$

would be

a. -0.075 V b. -0.15 V c. 0.075 V d. 0.15 V e. 0.30 V

14.39 If the value of -0.50 V were assigned to the standard hydrogen half cell instead of the currently accepted value of zero, the cell potential of a given electrochemical cell would

a. increase by 0.50 V. b. decrease by 0.50 V.

c. increase by 1.0 V. d. decrease by 1.0 V. e. remain unchanged.

14.40 Given the standard reduction potentials

$$Cr^{3+} + 3e \rightarrow Cr \quad -0.74 \text{ volts}$$

$$Pb^{2+} + 2e \rightarrow Pb \quad -0.13 \text{ volts}$$

What is the standard cell potential, E°, for the reaction

$$2\ Cr + 3\ Pb^{2+} \rightarrow 2\ Cr^{3+} + 3\ Pb$$

a. -0.87 volts b. 1.09 volts c. 0.61 volts d. -1.87 volts

e. None of the above.

64% 14.37 (d) $Mg^{2+} + 2e \rightarrow Mg$

Chlorine is obtained by electrolytic oxidation. Chlorine is used to oxidize bromide ion.

96% 14.38 (d) $2\ H^+ + 2\ e \rightarrow H_2$ at 1.00 M H^+ and 1.00 atm H_2 is the reference half-cell and it has been assigned the potential of 0.00 volt.

32% 14.39 (e) The cell potential of a complete reaction is independent of any value assigned to the reference half-cell. The potential is related to the difference in energy of reactants and the products.

14.40 (c)

$2(Cr \rightarrow Cr^{3+} + 3\ e)$	0.74 V
$3(Pb^{2+} + 2\ e \rightarrow Pb)$	-0.13 V
$2\ Cr + 3\ Pb^{2+} \rightarrow 2\ Cr^{3+} + 3\ Pb$	0.61 V

The half-reaction table below is to be used for questions 14.41 and 14.42.

Half Reaction	E° (volts)
$Mn^{2+} + 2e \rightarrow Mn$	-1.03
$Sn^{2+} + 2e \rightarrow Sn$	-0.14
$Hg_2Cl_2 + 2e \rightarrow 2Hg + 2Cl^-$	0.27
$Cu^+ + e \rightarrow Cu$	0.52
$Ag^+ + e \rightarrow Ag$	0.80
$Br_2 + 2e^- \rightarrow 2Br^-$	1.09

14.41 The strongest reducing agent of this series is

a. Mn^{2+} b. Br_2 c. Br^- d. Sn e. Mn

14.42 The element or ion which will oxidize mercury but will not oxidize bromide is

a. Mn b. Cu^+ c. Mn^{2+} d. Ag e. Cu

The half-reaction table below is to be used for questions 14.43-14.46.

Half Reaction	E° (volts)
$Br_2 + 2e \rightarrow 2Br^-$	1.09
$Hg_2^{2+} + 2e \rightarrow 2Hg$	0.80
$Cu^{2+} + 2e \rightarrow Cu$	0.34
$2H^+ + 2e \rightarrow H_2$	0.00
$Sn^{2+} + 2e \rightarrow Sn$	-0.14
$Fe^{2+} + 2e \rightarrow Fe$	-0.41
$Al^{3+} + 3e \rightarrow Al$	-1.67

88% 14.41 (e) Only three reducing agents, Br^-, Sn and Mn, are among the five responses.

$Mn \rightarrow Mn^{2+} + 2\ e \quad E° = 1.03$

The strongest reducing agent has the largest positive half-cell potential.

76% 14.42 (b) In this problem only two oxidizing agents, Cu^+ and Mn^{2+}, are given.

$2\ Cu^+ + 2e \rightarrow 2Cu$	0.52 V
$2Hg + 2Cl^- \rightarrow Hg_2Cl_2 + 2e$	-0.27 V
$2Cu^+ + 2Hg + 2Cl^- \rightarrow 2Cu + Hg_2Cl_2$	0.25 V

This reaction has a positive cell potential so it will go as it is written.

$2Cu^+ + 2e \rightarrow 2Cu$	0.52 V
$2Br^- \rightarrow Br_2 + 2e$	-1.09 V
$2Cu^+ + 2Br^- \rightarrow 2Cu + Br_2$	-0.57 V

This reaction has a negative potential so it will not go as it is written.

14.43 An element or ion which will oxidize iron but will not oxidize mercury is

a. Al^{3+} b. Sn^{2+} c. Br^- d. Cu e. Al

14.44 The strongest reducing agent of this series is

a. H_2 b. Br_2 c. Br^- d. Al e. Al^{3+}

14.45 An element which would dissolve in HNO_3 without the evolution of H_2 would be

a. Fe b. Sn c. Cu d. Al

14.46 From a consideration of the standard potentials, which of the following would give "galvanic protection" to iron: Br_2, Cu, Hg, Al and Sn.

a. Br_2 b. Cu c. Hg d. Al e. Sn

The half-reaction table below is to be used for questions 14.47 and 14.48.

Half Reactions	E° (volts)
$Hg_2^{2+} + 2e \rightarrow 2Hg$	0.80
$Sn^{2+} + 2e \rightarrow Sn$	-0.14
$Mg^{2+} + 2e \rightarrow Mg$	-2.38
$Br_2 + 2e \rightarrow 2Br^-$	1.09
$Cu^{2+} + 2e \rightarrow Cu$	0.34
$Fe^{3+} + e \rightarrow Fe^{2+}$	0.77
$Zn^{2+} + 2e \rightarrow Zn$	-0.76

14.47 The strongest oxidizing agent in this series is

a. Mg^{2+} b. Mg c. Br_2 d. Br^- e. Hg_2^{2+}

14.48 The strongest reducing agent is

a. Zn b. Hg_2^{2+} c. Br_2 d. Hg e. Mg

68% 14.43 (b) The oxidizing agents are Al^{3+} and Sn^{2+}.

$Sn^{2+} + 2e \rightarrow Sn$	-0.14 V	$Sn^{2+} + 2e \rightarrow Sn$	-0.14 V
$Fe \rightarrow Fe^{2+} + 2e$	0.41 V	$Hg \rightarrow Hg_2^{2+} + 2e$	-0.80 V
$Sn^{2+} + Fe \rightarrow Sn + Fe^{2+}$	0.27 V	$Sn^{2+} + Hg \rightarrow Sn + Hg_2^{2+}$	-0.94 V

48% 14.44 (d) The reducing agents are H_2, Br^- and Al.

$Al \rightarrow Al^{3+} + 3e$ 1.67 V

The half reaction written with the species functioning as a reducing agent that has the largest positive potential is the strongest reducing agent.

80% 14.45 (c) Fe, Sn and Al will all displace H_2 from water.

Cu will not displace H_2.

$Cu \rightarrow Cu^{2+} + 2e$	-0.34 V
$2H^+ + 2e \rightarrow H_2$	0.00 V
$Cu + 2H^+ \rightarrow Cu^{2+} + H_2$	-0.34 V

A reaction of HNO_3 acid with copper is represented by the equation below.

$$3Cu + 8H^+ + 2NO_3^- \rightarrow 3Cu^{2+} + 2NO + 4H_2O$$

84% 14.46 (d) Which element will reduce Fe^{2+} to Fe?

$2(Al \rightarrow Al^{3+} + 3e)$	1.67 V
$3(Fe^{2+} + 2e \rightarrow Fe)$	-0.41 V

14.47 (c) Bromine has the greatest attraction for electrons as the large positive E° value indicates.

14.48 (e) $Mg \rightarrow Mg^{2+} + 2e$ E° = 2.38

Magnesium is the best reducing agent as it has the largest E° potential when the half-reactions in the table are reversed.

The half-reaction table below is to be used for questions 14.49-14.51.

Half Reactions	E° (volts)
$Au^{3+} + 3e \rightarrow Au$	1.50
$Ag^{+} + e \rightarrow Ag$	0.80
$Cu^{2+} + 2e \rightarrow Cu$	0.34
$Pb^{2+} + 2e \rightarrow Pb$	-0.13
$Ni^{2+} + 2e \rightarrow Ni$	-0.25
$Al^{3+} + 3e \rightarrow Al$	-1.66

14.49 The strongest oxidizing agent of this series is

a. Al^{3+} b. Al c. Au^{3+} d. Au

14.50 The strongest reducing agent of this series is

a. Al^{3+} b. Al c. Au^{3+} d. Au

14.51 A species which would react with an acid to produce hydrogen and would also reduce Cu^{2+} is

a. Au b. Ag c. Pb^{2+} d. Al^{3+} e. Ni

14.52 If all of the species below are in their standard states, which is the strongest oxidizing agent?

a. H^{+} b. K c. F^{-} d. Al^{3+} e. Pb^{2+}

14.53 Protection of iron from corrosion can be accomplished by making an electrical contact between iron and certain other metals. Metal(s) which would provide protection is (are)

1. Hg 2. Sn 3. Al 4. Ag

a. 1 only b. 3 only c. 1 and 2 only d. 2 and 4 only

e. 3 and 4 only

76% 14.49 (c) There are two oxidizing agents listed, Al^{3+} and Au^{3+}. Au^{3+} has a more positive potential.

$Au^{3+} + 3e \rightarrow Au \qquad E° = 1.50$ V

80% 14.50 (b) Two reducing agents are listed, Al and Au.

$Al \rightarrow Al^{3+} + 3e \qquad E° = 1.66$ V.

Al has a more positive potential.

76% 14.51 (e) For this question there are only three reducing agents, Au, Ag and Ni, to consider.

$Ni + 2H^{+} \rightarrow Ni^{2+} + H_2, \qquad E° = 0.25$ V

$Ni + Cu^{2+} \rightarrow Ni^{2+} + Cu, \qquad E° = 0.59$ V

40% 14.52 (a) There are three oxidizing agents to consider (H^{+}, Al^{3+}, and Pb^{2+}).

	E°
$2\,H^{+} + 2e \rightarrow H_2$	0.00 V
$Pb^{2+} + 2e \rightarrow Pb$	-0.13 V
$Al^{3+} + 3e \rightarrow Al$	-1.66 V

There is a more positive potential for the H^{+}, H_2 half-cell. It is helpful for a student to memorize an abbreviated electromotive series.

80% 14.53 (b) Aluminum is the only element listed that is a stronger reducing agent than iron. The iron in a "tin can" rusts faster because the tin facilitates the oxidation of the iron.

14.54 A piece of iron half immersed in a sodium chloride solution will corrode more rapidly than a piece of iron half immersed in pure water because

a. the sodium ions oxidize the iron atoms.
b. the chloride ions oxidize the iron atoms.
c. the chloride ions form a precipitate with iron.
d. the chloride ions increase the pH of the solution.
e. the sodium ions and chloride ions carry a current through the solution.

Nernst Equation

The information below is for questions 14.55 and 14.56.
For cell reaction, $3Hg + Cr_2O_7^{2-} + 14H^+ \rightarrow 2Cr^{3+} + 3Hg^{2+} + 7H_2O$, the standard cell potential is 0.48 V.

14.55 To determine the cell potential at nonstandard conditions, the value that should be used for n in the Nernst equation is

a. 1 b. 2 c. 3 d. 4 e. 6

14.56 Which change(s) will result in an increase in the cell potential?

1. Increasing the dichromate ion concentration.
2. Increasing the pH.
3. Increasing the mercury ion concentration.
4. Increasing the amount of mercury.

a. 1 only b. 2 only c. 3 only d. 1 and 4 only
e. 2 and 4 only

14.57 From a consideration of the Nernst equation, $E = E° - \frac{0.0592}{n} \log Q$, when the concentration of the reactants and products are all 1.0 molar then

a. E = 1.0 b. E = E° c. E = 0 d. E° = 0 e. $K_c = Q$

32% 14.54 (e) More students selected response (b) than (e)! Chloride ioins can only function as a reducing agent. The sodium chloride acts as an electrolyte. Cars rust more rapidly in areas where roads are salted during the winter. Why?

14.55 (e)

$$\begin{array}{l} 14\ H^+ + Cr_2O_7^{2-} + 6\ e \rightarrow 2\ Cr^{3+} + 7\ H_2O \\ \underline{3(Hg \rightarrow Hg^{2+} + 2e)} \end{array}$$

There is a 6 electron change in the balanced half reactions.

28% 14.56 (a) By increasing the pH we are lowering the H^+ concentration. More students selected response (d) than (a). Increasing the amount of Hg does affect the potential because we haven't changed the concentration of the Hg as it is a pure phase.

14.57 (b) $E = E° - \frac{0.059}{n} \log \frac{1}{1} = E°$

14.58 For a certain oxidation-reduction reaction, E° is negative. This means that

a. $\Delta G°$ is negative and K is less than 1.

b. $\Delta G°$ is negative and K is greater than 1.

c. $\Delta G°$ is zero and K is greater than 1.

d. $\Delta G°$ is positive and K is greater than 1.

e. $\Delta G°$ is positive and K is less than 1.

14.59 A cell consists of a magnesium electrode immersed in a solution of magnesium chloride and a silver electrode immersed in a solution of silver nitrate. The two half cells are connected by means of a salt bridge. It is possible to increase the voltage of the cell by

a. addition of sodium chloride to both half cells.

b. increasing the Mg^{2+} concentration and decreasing the Ag^{+} concentration.

c. increasing the size of the Mg electrode and decreasing the size of the Ag electrode.

d. decreasing the size of the Mg electrode and increasing the size of the Ag electrode.

e. decreasing the Mg^{2+} concentration.

14.60 The cell potential of a new alkaline dry cell is 1.5 V. If, in a particular battery, the species involved in the cell reaction were at equilibrium, the cell potential would be

a. more than 1.5 V. b. 1.5 V. c. between 0 V and 1.5 V.

d. 0 V. e. cannot say unless the temperature is given.

60% 14.58 (e) If E° is negative then ΔG° is positive since $\Delta G^\circ = \mathcal{F}{-n}\ E^\circ$ ($-x- = +$). If E° is negative then K must be less than one since

$$K = 10^{nE^\circ/0.059}$$

$$10^{+b} = \text{a number} > 1$$
$$10^\circ = 1$$
$$10^{-b} = \text{a number} < 1$$

14.59 (e) $Mg + 2\ Ag^+ \rightleftarrows Mg^{2+} + 2\ Ag$

By applying LeChatelier's principle, we can predict the effect that a change in concentration will have on the cell potential (or use the Nernst equation).

(a) Decrease because Ag^+ will precipitate as AgCl.

(b) This will shift the equilibrium to the left thus decreasing the cell potential.

(c) No effect.

(d) No effect.

76% 14.60 (d) $\underline{E}$ of a cell represents the driving force of a chemical reaction. At equilibrium there is no driving force and $\underline{E} = 0$.

14.61 For the cell, $Zn(s) \mid Zn^{2+} \parallel Ag^{+} \mid Ag(s)$, the standard cell potential is 1.56 V. A cell using these reagents was made and the observed potential was 1.35 V. A possible explanation for the observed voltage is

a. There was 0.0100 mole of Zn^{2+} and 0.200 mole of Ag^{+}.
b. The $[Zn^{2+}]$ was larger than $[Ag^{+}]$.
c. The Ag electrode was twice as large as the Zn electrode.
d. The volume of the Zn^{2+} solution was larger than the volume of the Ag^{+} solution.
e. The volume of the Ag^{+} solution was larger than the volume of the Zn^{2+} solution.

14.62 A concentration cell is set up in which the cathode compartment consists of silver wire dipped in 1.0 M Ag^{+} and the anode compartment consists of a silver wire dipped in a saturated silver chloride solution. Which of the factor(s) below is (are) true for this cell?

1. $E = E°$ 2. $E° = 0$ 3. $E > 0$

a. 1 only   b. 2 only   c. 3 only   d. 1 and 2 only
e. 2 and 3 only

14.63 For a cell which involves iron as the anode, Fe, Fe^{2+} (0.00100 M) and Ag, Ag^{+} (1.50 M), what is the value of the reaction quotient, Q, for the cell?

a. 4.44×10^{-4} b. 6.67×10^{-4} c. 6.67×10^{-3} d. 1500
e. 2250

72% 14.61 (b) $Zn + 2Ag^+ \rightarrow Zn^{2+} + 2Ag$

If $[Ag^+] = [Zn^{2+}] = 1.0$ M then $E = E° = 1.56$ V. If Zn^{2+} is larger than Ag^+ then $E < E°$.

14.62 (e) $Ag + Ag^+ \rightarrow Ag^+ + Ag \qquad E° = 0.00$

A potential is measured for this cell because of the concentration difference, therefore, $E > 0$. $E = E°$ at equilibrium, but this cell is not at equilibrium so E does not equal zero.

44% 14.63 (a) $$E = E° - \frac{0.059}{n} \log \frac{[Fe^{2+}]}{[Ag^+]^2}$$

$$Q = \frac{[0.001]}{[1.50]^2} = 4.44 \times 10^{-4}$$

Questions 14.64 - 14.66 refer to the following:

A voltaic cell is made by placing a copper electrode into a compartment with the Cu^{2+} concentration equal to 1.0×10^{-4} M and a Pt electrode into the other compartment dipping into a solution that is 1.0×10^{-3} M Fe^{3+} and 1.0×10^{-2} M Fe^{2+}.

Half Reactions	E° (volts)
$Fe^{3+} + e \rightarrow Fe^{2+}$	0.77
$Cu^{2+} + 2e \rightarrow Cu$	0.34

14.64 What is the value of E° for this cell and which electrode is the anode?

a. 0.43, Cu b. 0.43, Pt c. 1.11, Cu d. 1.11, Pt

e. 1.20, Pt

14.65 What is E for the cell at the concentrations given above?

a. 0.37 b. 0.49 c. 0.52 d. 1.05 e. 1.17

14.66 What is the equilibrium constant for the spontaneous cell reaction above?

a. 1.9×10^{7} b. 3.8×10^{14} c. 4.1×10^{16} d. 6.5×10^{18}

e. 4.2×10^{37}

14.67 The ability of all the species below to function as oxidizing agents would be affected by the hydronium ion concentration EXCEPT

a. Cl_2 b. NO_3^- c. MnO_4^- d. NO_2^- e. CrO_4^{2-}

14.64 (a) $2\,(Fe^{3+} + e \rightarrow Fe^{2+})$ 0.77 V cathode

$Cu \rightarrow Cu^{2+} + 2\,e$ -0.34 V anode

$2\,Fe^{3+} + Cu \rightarrow Cu^{2+} = 2Fe^{2+}$ 0.43 V

14.65 (b) $E = E° - \frac{0.059}{2} \log \frac{[1.0 \times 10^{-2}]^2[1.0 \times 10^{-4}]}{[1.0 \times 10^{-3}]^2}$

$E = 0.43 - 0.030 \log (1.0 \times 10^{-2})$

$E = 0.43 - 0.030\,(-2) = 0.49$ V

14.66 (b) $0 = 0.43 - \frac{0.059}{2} \log K$

$\log K = \frac{2(-0.43)}{-0.059} = 14.58$

$K = 3.8 \times 10^{14}$

14.67 (a) $Cl_2 + 2\,e \rightarrow 2\,Cl^-$

The hydronium ion does not appear as a reactant or product in the balanced half-reaction above, so a change in concentration will have no effect. In contrast, consider one of the other oxidizing agents:

$MnO_4^- + 8\,H^+ + 5\,e \rightarrow Mn^{2+} + 4\,H_2O$

It can be seen that H^+ is a reactant in the half-reaction above.

REPRESENTATIVE METALS

15.1 A group of metals which all melt below 200°C is

a. the alkali metals.

b. the alkaline earth metals.

c. the transition metals.

d. the lanthanides.

e. the actinides.

15.2 All of the following statements are true EXCEPT

a. Most metal oxides are basic.

b. Most metals have positive reduction potentials.

c. Most metals are dense solids at 500 K.

d. Metals are good heat conductors.

e. Metals are good electrical conductors.

15.3

Group I A	Melting Point °C
Li	179
Na	98
K	64
Rb	39
Cs	28

The trend in melting points of the Group I A elements is best explained by

a. a decrease in van der Waals forces with increasing atomic number.

b. a decrease in metallic bonding interactions due to an increase in the size of the atom.

c. a decrease in the ionization energy with increasing atomic number.

d. a decrease in the electron affinity with increasing atomic number.

e. a decrease in the electronegativity with increasing atomic number.

15.4 Which of the following metals does not react with acids to give H_2?

a. Cu b. Fe c. Al d. Li e. Zn

76% 15.1 (a) The alkali metals are very low melting solids. Let us compare alkali metals and the alkaline earths.

	Melting Point, °C		Melting Point, °C
Li	179	Be	1283
Na	98	Mg	656
K	64	Ca	850
Rb	39	Sr	770
Cs	28	Ba	710

The transition metals have higher melting points than the alkaline earths.

64% 15.2 (b) Most metals are above hydrogen in the electromotive series. For example, consider zinc ion-zinc half-cell written as a reduction half-reaction

$$Zn^{2+} + 2\ e \rightarrow Zn,\ E° = -0.76\ V$$

Most metals have negative reduction potentials.

15.3 (b) This question must be explained on the basis that these elements are metals, so we must consider metallic bonding interactions.

15.4 (a) Which metal is below hydrogen in the electromotive series? Some other metals besides copper which fall below hydrogen are Au, Pt, Ag and Hg.

15.5 When magnesium burns in air, the combustion product(s) is(are)

a. a mixture of magnesium oxide and a magnesium superoxide.

b. an ionic solid containing O^{2-} ions and N^{3-} ions.

c. a mixture of Mg_3O, MgO_2 and MgN.

d. a mixture of Mg_2O_2, Mg_3N_2.

e. a mixture of MgO, MgO_2 and Mg_3N_2.

15.6

1. $K(s) + O_2(g) \rightarrow KO_2(s)$
2. $2K(s) + O_2(g) \rightarrow K_2O_2(s)$
3. $4K(s) + O_2(g) \rightarrow K_2O(s)$

Which of the equation(s) above represents actual reactions of potassium with oxygen?

a. 1 only b. 2 only c. 3 only d. 2 and 3 only

e. 1, 2, and 3

15.7 Sodium oxide (Na_2O) is added to excess water. Which one of the following would occur?

a. Reaction to yield an aqueous solution of sodium hydroxide.

b. Reaction to yield hydrogen and an aqueous solution of sodium hydroxide.

c. Reaction to yield oxygen and an aqueous solution of sodium hydroxide.

d. Reaction to yield oxygen, hydrogen and an aqueous solution of sodium hydroxide.

e. No reaction will occur.

15.8 Which of the following has the highest ionization energy for the removal of the <u>third</u> electron?

a. Mg b. Al c. Si d. Cl e. Ar

30% 15.5 (b) The products are MgO and Mg_3N_2. No peroxides or superoxides ae formed with Mg. Superoxides and peroxides are formed between some of the alkali metals and oxygen.

15.6 (e) Potassium forms a mixture of products when it reacts with oxygen. An oxide, peroxide and a superoxide are formed.

76% 15.7 (a) Na_2O (metal oxides) is a basic anhydride. There is an acid-base reaction with water.

$$Na_2O(s) + H_2O \rightarrow 2\ Na^+(aq) + 2\ OH^-(aq)$$

15.8 (a) The third electron that is removed from Mg is not a valence electron. It is coming from a lower energy level which requires a great deal of energy. All the other atoms have at least three valence electrons.

The information below is for questions 15.9 and 15.10.

I. The reaction of a metal with water is very vigorous.

II. The metal reacts with water evolving hydrogen.

III. The metal is in period 4.

IV. The metal reacts with water causing phenolphthalein in the water to change from colorless to pink.

15.9 The metal is acting as

a. a reducing agent. b. an oxidizing agent. c. an acid.

d. a base. e. an amphoteric substance.

15.10 The metal is probably the element

a. calcium b. magnesium c. sodium d. lithium e. potassium

15.11 Which of the following is initially obtained commercially by electrolytic reduction?

a. Chlorine b. Aluminum c. Copper d. Bromine e. Fluorine

15.12 Which of the following is not a characteristic property of alkaline earth elements in Group II A of the Periodic Table?

a. They form 2^+ ions.

b. Their oxides dissolve in water to form the hydroxides.

c. The ionic radius is smaller than the atomic radius.

d. The atomic radius decreases with increasing atomic number.

e. They are strongly electropositive elements.

15.13 Cryolite, Na_3AlF_6, is used in the production of

a. Na b. Al c. HF d. NaF e. F_2

15.14 All the following represent known compounds of potassium EXCEPT

a. K_2O b. K_2O_3 c. K_2O_2 d. KO_2 e. K_2SO_4

15.15 An element which has the same valence electron configuration as tin is

a. silicon b. strontium c. antimony d. scandium e. selenium

48% 15.9 (a) The oxidation number of hydrogen changes from 1+ to 0 so it is reduced. The metal does the reducing so it is the reducing agent.

80% 15.10 (e) Both potassium and calcium are in Period 4. The reaction of potassium with water is much more vigorous than the reaction of calcium with water.

56% 15.11 (b) Chlorine and fluorine are obtained by electrolytic oxidation. Copper is purified by electrolytic reduction, but it is obtained commercially by a smelting process or by iron displacing Cu^{2+} ions from solution.

15.12 (d) In a given family, the size of the atom increases from the top of the Periodic Table to the bottom of the table because of the addition of more electrons to higher energy levels.

72% 15.13 (b) The cryolite is used to help lower the melting point of the aluminum salts. Aluminum is obtained by electrolysis.

84% 15.14 (b) Potassium forms three known oxides, potassium oxide, peroxide, and superoxide.

96% 15.15 (a) Both have an outer electronic configuration of ns^2np^2. Tin and silicon belong to the same family.

15.16 The principle oxidation states of lead are 2+ and 4+. Consequently, the expected oxidation states for antimony are

a. 3+ and 5+ b. 1+ and 3+ c. 2+ and 3+ d. 2+ and 4+
e. 1+ and 5+

15.17 If the trends common in the Periodic Table are followed, what should be the most stable oxidation state of the element that would be located below bismuth, Bi?

a. +1 b. +2 c. +3 d. +5 e. -3

15.18 Calcium ions are removed from water and replaced by sodium ions in a home water softener. This is done by

a. osmosis. b. reverse osmosis. c. using an anion exchange column.
d. using a cation exchange column. e. precipitation.

15.19 What products are expected when potassium superoxide, KO_2, reacts with water?

a. KOH b. $KOH + O_2$ c. $K_2O + O_2$ d. $K_2O + H_2$ e. $KOH + K_2O + H_2$

15.20 Which one of the following ions will react (hydrolyze) to the greatest extent with water?

a. Li^+ b. Cs^+ c. Mg^{2+} d. Ba^{2+} e. Al^{3+}

15.21 Which one of the following compounds has amphoteric properties?

a. $Be(OH)_2$ b. $Mg(OH)_2$ c. $Ba(OH)_2$ d. CsOH e. LiOH

15.22 Which one of the following metals has the greatest abundance in the earth's crust?

a. sodium b. barium c. iron d. aluminum e. lithium

15.23 Which one of the following metals forms the most covalent like compounds?

a. Li b. Cs c. Be d. Mg e. Ba

68% 15.16 (a) Pb $6s^2\ 6p^2$ Sb $5s^2\ 5p^3$

Pb^{2+} $6s^2\ 6p^o$ Sb^{3+} $5s^2\ 5p^o$

Pb^{4+} $6s^o\ 6p^o$ Sb^{5+} $5s^o\ 5p^o$

36% 15.17 (c) The heavy metals by bismuth all have the most stable oxidation state as the group number -2. For example Pb^{2+} is more stable than Pb^{4+}; Ti^+ is more stable than Tl^{3+}; Bi^{3+} is more stable than Bi^{5+} (BiO_3^-).

15.18 (d)

15.19 (b) Metal oxides react with water to produce a basic solution so KOH(aq) must be one of the products. When the crystalline lattice of KO_2 is broken, oxygen is released as a product.

15.20 (e) $Al(H_2O)_6^{3+} + H_2O \rightleftharpoons Al(H_2O)_5OH^{2+} + H_3O^+$

Hydrolysis reactions are important reactions when a metal ion has a large charge/size ratio. The charge/size ratio is by far the largest for the Al^{3+} ion.

15.21 (a) Elements which are borderline in their metal and nonmetal properties have amphoteric properties. They can act as a base in the presence of an acid and an acid in the presence of a base.

$Be(OH)_2(s) + 2H^+(aq) \rightleftharpoons Be^{2+}(aq) + 2H_2O$

$Be(OH)_2(s) + 2OH^-(aq) \rightleftharpoons Be(OH)_4^{2-}(aq)$

15.22 (d) Aluminum is the most common metal in the earths crust. It occurs in nature in silicates such as micas and feldspars as the hydrooxide (bauxite) and as cryolite.

15.23 (c) Beryllium has a unique chemical behavior as compared to the other Group II elements. It is involved predominantly in covalent chemistry because the atom is very small and it has a relatively high ionization energy. It has a high charge/radius ratio.

15.24 Which one of the following oxides has the most basic character?

a. SiO_2 b. BeO c. BaO d. Al_2O_3 e. Ga_2O_3

15.25 Which one of the following ions has the greatest ability to polarize negative ions?

a. Li^+ b. Na^+ c. K^+ d. Rb^+ e. Cs^+

15.26 Which one of the following metals exhibits the inert s pair effect?

a. Na b. K c. Mg d. Al e. Sn

15.24 (c) Which metal ion has the lowest charge/radius ratio? Which metal has the lowest ionization energy?

84% 15.25 (a) Which metal ion has the largest charge/radius ratio? (Which metal ion is the smallest?)

52% 15.26 (e) Tin has the electronic configuration $5s^2\ 5p^2$. The electronic configuration for Sn^{2+} is $5s^2\ 5p^0$. The s electrons are rather difficult to remove. Sn^{2+} is quite stable and Sn^{4+} is a relatively strong oxidizing agent.

16

NONMETALS

Hydrogen

16.1 The most important use of elemental hydrogen is as

a. an acid. b. a base. c. an electrolyte. d. a reducing agent.
e. an oxidizing agent.

16.2 Hydrogen can be readily obtained by all the reactions below EXCEPT

a. $CH_4 + 2Cl_2 \rightarrow CCl_4 + 2H_2$
b. $2H_2O \xrightarrow{\text{electricity}} 2H_2 + O_2$
c. $CaH_2 + 2H_2O \rightarrow Ca(OH)_2 + 2H_2$
d. $Zn + 2HCl(aq) \rightarrow ZnCl_2(aq) + H_2$
e. $C(s) + H_2O(g) \xrightarrow{1500°C} CO(g) + H_2(g)$

16.3 All of the following are methods that can be used for the production of hydrogen EXCEPT

a. $Li + H_2O$ b. $MgH_2 + H_2O$ c. natural gas + H_2O
d. carbon + H_2O e. the atmosphere

16.4 Which of the following hydrides will react with water to produce hydrogen gas?

a. HF b. H_2S c. CaH_2 d. CH_4 e. NH_3

16.5 All of the following can be used as methods for producing binary hydrides EXCEPT

a. $H_2 + Cl_2 \xrightarrow{uv} 2HCl$
b. $Mg_3N_2 + 3CaH_2 \rightarrow 2NH_3 + 3Mg + 3Ca$
c. $P_4 + 3OH^- + 3H_2O \rightarrow PH_3 + 3H_2PO_2^-$
d. $NaCl + H_2SO_4 \rightarrow HCl + NaHSO_4$
e. $2H_2 + O_2 \rightarrow 2H_2O$

35% 16.1 (d) Hydrogen is used as a fuel (reducing agent) in the space program. It is used to reduce (hydrogenate) organic compounds. Hydrogen is used in large quantities to produce ammonia.

52% 16.2 (a) $CH_4 + 2Cl_2 \rightarrow CCl_4 + 4HCl$

16.3 (e) Hydrogen is present on the sun but it is too light of a molecule to exist in the earth's atmosphere.

16.4 (c) $CaH_2 + 2H_2O \rightarrow Ca(OH)_2 + 2H_2$

40% 16.5 (b) Ww should not expect that the combination of two salts would produce two very reactive elements. Magnesium and calcium are prepared by electrolysis.

Boron, Carbon and Silicon

16.6 Boron does each of the following EXCEPT

a. acts as a reducing agent with some metal oxides.

b. reduces concentrated nitric acid.

c. forms compounds with the halogens that can hydrolyze.

d. forms some covalent compounds that are Lewis bases.

e. forms hydrides that ignite spontaneously in air.

16.7 The number of valence electrons that are available for bonding in diborane is

a. 6 b. 8 c. 10 d. 12 e. 14

16.8 Which of the following statements concerning diborane is INCORRECT?

a. There are two three-center bonds.

b. It is known as an electron deficient compound.

c. It is flammable in air.

d. It has a boron-boron bond.

e. None of the above.

16.9 Which of the following statements concerning diborane is INCORRECT?

a. It is a Lewis base.

b. It has two bridging hydrogen atoms.

c. There is no rotation of the BH_2 group about the B - B axis.

d. It is spontaneously flammable in air.

e. Four of the boron hydrogen bonds each uses two electrons.

16.10 The carbon atoms in graphite are

a. diatomic molecules.

b. sheets with each atom attached to three other atoms.

c. at the corners of a tetrahedron.

d. long chains.

e. none of the above.

48% 16.6 (d) Boron compounds such as BCl_3 and B_2H_6 are electron deficient and function as Lewis acids.

40% 16.7 (d) B_2H_6

$$\begin{aligned} 2\text{ B} &= 2 \times 3 = 6\text{ e} \\ 6\text{ H} &= 6 \times 1 = \underline{6\text{ e}} \\ &\quad\quad\quad\ 12\text{ e} \end{aligned}$$

68% 16.8 (d)

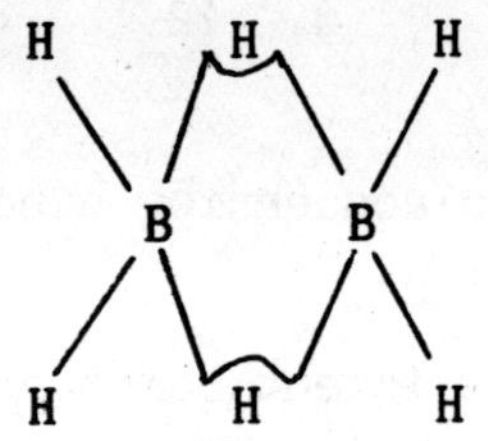

76% 16.9 (a) Diborane is an electron deficient compound so it will function as a Lewis acid.

16.10 (b) Graphite is a layered structure with each carbon atom bonded to three other atoms. In the diamond structure, each carbon is bonded to four other carbon atoms.

16.11 All of the following are important reducing agents EXCEPT

a. H_2 b. Fe   c. C   d. CO   e. CO_2

16.12 Silicon does not undergo extensive catenation (bonding to itself) because

1. the silicon-silicon bond is a weak bond.
2. silicon forms strong bonds with a variety of nonmetal atoms.
3. silicon has a very high melting point.

a. 1 only   b. 2 only   c. 3 only   d. 1 and 2   e. 1, 2, and 3

16.13 All of the following are well known molecules EXCEPT

a. C_2H_6 b. C_2H_4 c. C_2H_2 d. Si_2H_6 e. Si_2H_4

16.14 In which one of the following is carbon best described as sp^3 hybridized?

a. diamond   b. graphite   c. carbon monoxide   d. carbon dioxide

e. calcium carbonate

16.15 Stalagtites and stalagmites are primarily

a. Al_2O_3 b. SiO_2 c. $Ca(HCO_3)_2$ d. $CaSiO_3$ e. $CaCO_3$

16.16 Silane is

a. another name for silicone.   b. another name for quartz

c. SiO_2 d. SiH_4 e. Na_2SiO_3

16.17 Acetylene can be prepared by

a. $2C + H_2 \rightarrow C_2H_2$

b. $CaC_2 + 2H_2O \rightarrow C_2H_2 + Ca(OH)_2$

c. $2CO_2 + 2H_2O \rightarrow C_2H_2 + 3O_2 + H_2$

d. $C_2H_6 + O_2 \rightarrow C_2H_2 + 2H_2O$

e. $2CO + 2HCl \rightarrow C_2H_2 + Cl_2 + O_2$

36% 16.11 (e) Carbon dioxide cannot readily function as a reducing agent since carbon is its highest oxidation state.

76% 16.12 (d) The Si-Si bond is very weak and the Si-O bond is very strong.

84% 16.13 (e) The silicon atoms are relatively large so they don't readily form double bonds. There isn't efficient sidewise overlap of the p-orbitals.

16.14 (a) In diamond. each carbon is bonded to four other carbons so the carbon is sp^3 hybridized.

16.15 (e)

16.16 (d)

16.17 (b)

16.18 The anhydride of carbonic acid is

a. C_3O_2 b. CO c. CO_2 d. C_2O_3 e. $NaHCO_3$

16.19 Which properties describe silicon?

1. High melting point.
2. A diamond-like structure.
3. A semiconductor.

a. 1 only b. 2 only c. 3 only d. 1 and 2 only
e. 1, 2 and 3

16.20 Semiconductors usually have the following properties or characteristics EXCEPT

a. a defect in the crystal.
b. the electrical conductance increases with temperature.
c. like metals, the resistance of the semiconductor increases with temperature.
d. defects are produced by the introduction of foreign atoms.
e. can convert radiant energy to electrical energy.

Nitrogen Family

16.21 All of the following statements about the elements N, P, and As are true EXCEPT

a. the hydrides have the general formula AH_3.
b. the oxides have the general empirical formulas A_2O_3 and A_2O_5.
c. the fluorides have the general formulas AF_3 and AF_5.
d. the ionization energies of these elements decrease as their atomic numbers increase.
e. all the elements exhibit an oxidation state of -3.

16.22 Which compound gives photochemical smog a brownish color?

a. N_2O_3 b. N_2O c. NO d. NO_2 N_2O_5

16.18 (c) $CO_2 + H_2O \rightarrow H_2CO_3$

this is an acid-base reaction so there is no change in oxidation number.

16.19 (e)

48% 16.20 (c) Semiconductors become conductors at elevated temperatures.

16.21 (c) Nitrogen cannot form more than four bonds nor can it have more than eight electrons about it.

16.22 (d)

16.23 Ammonia has the following important uses EXCEPT

a. a readily available oxidizing agent.

b. production of ammonium nitrate.

c. production of nitrogen(II) oxide.

d. production of nitric acid.

e. all of the above.

16.24 Of the following compounds, which ones have nitrogen with a 5+ oxidation state?

1. O_2NONO_2 2. $HONO_2$ 3. ONF_3 4. O_2NNO_2

a. 1 and 2 only b. 3 and 4 only c. 2 and 4 only

d. 1 and 3 only e. 1, 2 and 3 only

16.25 Which pair of substances, when mixed, will generate ammonia gas?

1. $NH_4Cl(s)$ and $NaOH(aq)$

2. $Ca_3N_2(s)$ and $H_2O(\ell)$

3. $NH_4Cl(s)$ and $HCl(aq)$

4. $NO_2(aq)$ and $H_2O(\ell)$

a. 1 and 2 only b. 2 and 3 only c. 3 and 4 only

d. 1 and 3 only 3. 2, 3 and 4 only

16.26 Ammonia has a much higher boiling point than phosphine, PH_3. This is best explained by

a. stronger dispersion forces between NH_3 molecules than between PH_3 molecules.

b. stronger dipolar forces between NH_3 molecules than between PH_3 molecules.

c. N-H covalent bonds being stronger than P-H covalent bonds.

d. phosphine has a greater bond polarity than that in ammonia.

e. P-H covalent bonds being stronger than N-H covalent bonds.

16.23 (a) Ammonia is a fuel. It is a reducing agent as nitrogen is in its lowest oxidation state.

88% 16.24 (e) N in N_2O_4 has a 4+ oxidation number (2 y + 4 (-2) = 0).

48% 16.25 (a) (1) $NH_4^+ + OH^- \rightarrow NH_3 + H_2O$

(2) $Ca_3N_2(s) + 6\ H_2O \rightarrow 3\ Ca(OH)_2(s) + 2\ NH_3(g)$

(3) $NH_4^+ + H^+$ (an acid plus an acid-no reaction).

(4) $3\ NO_2(g) + H_2O \rightarrow 2\ HNO_3(aq) + NO$

64% 16.26 (b) Hydrogen bonding is a very strong dipole-dipole interaction. Boiling points are related to the strength of the intermolecular forces.

16.27 Write the equation for the oxidation of gaseous ammonia with oxygen in the presence of platinum as a catalyst to produce NO. The sum of all the coefficients in the balanced equation is

a. 15 b. 17 c. 19 d. 21 e. 23

16.28 The commercial preparation of nitric acid involves a number of reactions. Which of the reaction sequences below summarizes part of this process?

a. $N_2 \xrightarrow{O_2} NO_2 \xrightarrow{N_2} NO$ b. $N_2 \xrightarrow{O_2} NO \xrightarrow{O_2} NO_2$

c. $NH_3 \xrightarrow{O_2} NO_2 \xrightarrow{N_2} NO$ d. $NH_3 \xrightarrow{O_2} N_2 \xrightarrow{O_2} NO$

e. $NH_3 \xrightarrow{O_2} NO \xrightarrow{O_2} NO_2$

16.29 Nitric acid is a powerful oxidizing agent. Which of the following substances is least likely to be produced when nitric acid oxidizes a metal?

a. NO b. NO_2 c. N_2O_4 d. N_2O_5 e. NH_4^+

16.30 Which of the following nonmetal oxides is the acidic anhydride of the acid with which it is paired?

a. NO_2 and HNO_2 b. NO_2 and HNO_3 c. NO and HNO_2
d. N_2O_3 and HNO_2 e. N_2O_4 and HNO_3

16.31 Which of the following nonmetal oxides is paired with the acid of which it is the anhydride?

a. NO_2 and HNO_3 b. N_2O_3 and HNO_2 c. CO and H_2CO_3
d. SO_2 and H_2SO_4 c. P_4O_{10} and H_3PO_3

16.32 The central atom(s) in all of the following species have intermediate oxidation states and can undergo self oxidation and reduction EXCEPT

a. NO_2^- b. NO_2 c. H_2O_2 d. O_2^- e. NO_3^-

16.27 (c) $4NH_3 + 5O_2 \rightarrow 4NO + 6H_2O$

48% 16.28 (e) See the answer to question 16.27 above.

56% 16.29 (d) An oxidizing agent is reduced. Therefore, N_2O_5 cannot be a reduction product because N in N_2O_5 is a 5+ oxidation number which is the same as the oxidation number in HNO_3.

96% 16.30 (d) The central atom in the anhydride and in the acid must have the same oxidation number.

$$N_2O_3 + H_2O \rightarrow 2\ HNO_2$$

16.31 (b) The central atom in the anhydride and in the acid must have the same oxidation number.

16.32 (e) N in the NO_3^- is in its highest oxidation (5+) and this corresponds to Group 5. We would expect it to act only as an oxidizing agent.

16.33 Which of the following diatomic molecules has the greatest bond strength?

a. F_2 b. HF c. HCl d. O_2 e. N_2

16.34 The reaction of magnesium nitride, Mg_3N_2, with water produces ammonia. Write a balanced molecular equation for the reaction. The sum of all the coefficients in the balanced equation is

a. 8 b. 9 c. 10 d. 11 e. 12

16.35 All of the following statements about the uses of chemical compounds are correct EXCEPT

a. HNO_3 is used in preparation of TNT.
b. HNO_3 is used as an oxidizing agent.
c. HNO_3 is used to produce some fertilizers.
d. NH_3 is used to produce HNO_3.
e. NH_3 is used as an oxidizing agent.

16.36 Nitric acid has the following important uses EXCEPT

a. production of ammonium nitrate.
b. scavenging of rust from iron.
c. oxidation of cyclohexane to adipic acid.
d. production of TNT.
e. all of the above.

16.37 Phosphoric acid could be prepared by the addition of water to all of the following compounds EXCEPT

a. PCl_5 b. $POCl_3$ c. $PCl_4{}^+PCl_6{}^-$ d. P_4O_6 e. PBr_5

16.38 Write the molecular equation for the hydrolysis of PCl_3 in excess H_2O. What is the sum of the coefficients?

a. 4 b. 5 c. 6 d. 7 e. 8

16.39 Phosphorous acid would be prepared by the addition of water to all of the following compounds EXCEPT

a. PCl_3 b. PBr_3 c. PI_3 d. $POCl_3$ e. P_4O_6

32% 16.33 (e) Remember N_2 has a triple bond.

$$:N\equiv N:$$

The H-F bond is very strong but a triple bond is much stronger.

16.34 (e) $Mg_3N_2 + 6H_2O \rightarrow 3Mg(OH)_2 + 2NH_3$

16.35 (e) NH_3 is a reducing agent. See question 16.27.

52% 16.36 (b) Hydrochloric acid is used to remove rust from iron. Nitric acid because it is an oxidizing agent would promote the formation of Fe_2O_3.

64% 16.37 (d) The oxidation number of phosphorus in the reactant must be the same as that in the product.

$$P_4O_6 + 6\ H_2O \rightarrow 4\ H_3PO_3$$

40% 16.38 (e) $PCl_3 + 3H_2O \rightarrow H_3PO_3 + 3HCl$

60% 16.39 (d) The phosphorus atom in phosphorous acid, H_3PO_3, has an oxidation number of 3+. Therefore, the reactants should also have P in a 3+ state since the preparations of H_3PO_3 which are given involve acid-base reactions. In $POCl_3$, P has an oxidation number of 5+.

16.40 What charge does the ion PF_6^n have?

a. n = -3 b. n = -1 c. n = +1 d. n = +3 e. n = +5

16.41 All of the following can be used as methods for producing hydrides EXCEPT

a. $N_2 + 3H_2 \xrightarrow{cat.} 2NH_3$

b. $Ca_3N_2 + 6H_2O \rightarrow 3Ca(OH)_2 + 2NH_3$

c. $P_4 + 6H_2 \rightarrow 4PH_3$

d. $P_4 + 3OH^- + 3H_2O \rightarrow PH_3 + 3H_2PO_2^-$

e. $POCl_3 + 3H_2O \rightarrow H_3PO_4 + 3HCl$

Oxygen Family

16.42 All of the following are important reducing agents EXCEPT

a. H_2 b. C c. CO d. O_2 e. Fe

16.43 Which of the following compounds is a strong oxidizing agent?

a. HF b. CaH_2 c. H_2O d. KI e. H_2O_2

16.44 The gas in the upper atmosphere that absorbs uv radiation is

a. H_2O b. He c. N_2 d. O_3 e. NO

16.45 The principal commercial source material for preparing oxygen is

a. the atmosphere.

b. electrolysis of H_2O.

c. thermal decomposition of $KClO_3$.

d. catalytic decomposition of H_2O_2.

e. coal plus water and heat.

80% 16.40 (b) P is in its highest oxidation state which corresponds to its group number.

$\therefore +5 + 6(-1) = -1$

16.41 (c) Most binary hydrides can be prepared by combining the appropriate elements. Phosphine, PH_3, is an exception to this rule. Why? We don't know.

16.42 (d) Oxygen is an oxidizing agent not a reducing agent.

16.43 (e) HF functions primarily as an acid. CaH_2 and KI are reducing agents.

16.44 (d) There is presently a great deal of concern that the O_3 "layer" may be depleted by reaction with pollutants such as the Freons.

16.45 (a) Oxygen can be obtained from the electrolysis of water but that is very expensive.

16.46 The radii of the species O, O^+ and O^- decrease in the following order:

a. $O>O^+>O^-$ b. $O>O^->O^+$ c. $O^+>O^->O$ d. $O^->O>O^+$ e. $O^->O^+>O$

16.47 All of the following species are paramagnetic EXCEPT

a. Ca^+ b. O_2 c. O_3 d. NO_2 e. Cl

16.48 Hydrogen peroxide functions as an oxidizing agent in all the following reactions EXCEPT

a. $3H_2O_2 + Cr_2O_7^{2-} + 8H^+ \rightarrow 2Cr^{3+} + 3O_2 + 7H_2O$

b. $2H_2O_2 \rightarrow 2H_2O + O_2$

c. $H_2O_2 + 2H^+ + 2I^- \rightarrow I_2 + 2H_2O$

d. $H_2O_2 + SO_3^{2-} \rightarrow SO_4^{2-} + H_2O$

e. $H_2O_2 + 2Fe^{2+} + 2H^+ \rightarrow 2Fe^{3+} + 2H_2O$

16.49 Which of the following statements is true?

a. Metal oxides are acidic anhydrides.

b. BaO_2 is barium superoxide.

c. Hydrogen peroxide can only function as an oxidizing agent.

d. An amphoteric substance can act as a base in the presence of an acid.

e. Nonmetal oxides react with water to form basic solutions.

16.50 Sulfur can exist in several structural forms at room temperature. These forms are known as

a. isobars b. azeotropes c. isomers d. isotopes e. allotropes

16.51 When sulfur burns in air, the combustion product is

a. an ionic oxide.

b. a gas which dissolves in water to produce sulfuric acid.

c. a gas which has a "rotten egg" odor.

d. a gas which dissolves in water to produce sulfurous acid.

e. a mixture of gases which react with water to produce both a strong acid and a weak acid.

16.46 (d) O^- is the largest (the most electrons per 8 protons).
O^+ is the smallest (the fewest electrons per 8 protons).

16.47 (c) All molecules and atoms that have an odd number of electrons are paramagnetic. The structure of O_3 is isoelectronic with NO_2^-.

16.48 (a) The dichromate ion oxidizes hydrogen peroxide (so it is functioning as a reducing agent) to oxygen.

16.49 (d) An amphoteric compound can act as an acid in the presence of a base and as a base in the presence of an acid.

92% 16.50 (e) Three different allotropes of sulfur are monoclinic sulfur, rhombic sulfur and plastic sulfur.

28% 16.51 (d) $S + O_2 \rightarrow SO_2$

$SO_2 + H_2O \rightarrow H_2SO_3$

In order to oxidize sulfur to SO_3, a catalyst is required.

16.52 All of the oxides given below would be expected to react with oxygen EXCEPT

a. P_4O_6 b. NO c. N_2O d. SO_2 e. SO_3

16.53 Which oxide when added to water produces the most acidic solution?

a. CO_2 b. CO c. Li_2O d. SO_2 e. SO_3

16.54 Which of the following compounds is most likely to act as a reducing agent?

a. H_3PO_4 b. H_2SO_3 c. Mn_2O_7 d. SiO_2 e. CO_2

16.55 Which one of the following elements is most likely to form a compound of the formula H_2XO_3 where X is

a. Se b. P c. As d. Sb e. Cl

16.56 Which of the following compounds is most likely to act as a reducing agent?

a. HNO_3 b. H_2SO_4 c. H_2S d. Cl_2 e. H_3PO_4

16.57 Sulfuric acid is a satisfactory reagent for all the following reactions EXCEPT

a. $2NaNO_2 + H_2SO_4 \rightarrow Na_2SO_4 + 2HNO_2$

b. $2NaI + H_2SO_4 \rightarrow Na_2SO_4 + 2HI$

c. $2NaNO_3 + H_2SO_4 \rightarrow Na_2SO_4 + 2HNO_3$

d. $CaF_2 + H_2SO_4 \rightarrow CaSO_4 + 2HF$

e. $2NH_3 + H_2SO_4 \rightarrow (NH_4)_2SO_4$

16.58 Which one of the following would be most likely to bring out the oxidizing properties of H_2SO_4?

a. F^- b. Cl^- c. O^{2-} d. Mg^{2+} e. Se^{2-}

80% 16.52 (e) In SO_3, sulfur is already in its highest oxidation state so we wouldn't expect further oxidation.

16.53 (e) $SO_3 + H_2O \rightarrow H_2SO_4$

SO_2 and CO_2 react with water to produce weak acids.

68% 16.54 (b) In H_3PO_4, Mn_2O_7, SiO_2 and CO_2, the central atom is in its highest oxidation state so these compounds (in particular Mn_2O_7) can function as oxidizing agents. Sulfur in H_2SO_3 is in an intermediate oxidation state.

16.55 (a) We need an element that can have a 4+ oxidation number. Look for an element from an even number family.

60% 16.56 (c) Look for a central atom in a low oxidation state. In HNO_3, H_2SO_4 and H_3PO_4, the central atom is in its highest oxidation state.

16.57 (b) Need to use H_3PO_4 because H_2SO_4 is a strong enough oxidizing agent to oxidize I^- to I_2.

30% 16.58 (e) Look for an ion that doesn't have a strong attraction for electrons.

Halogens

16.59 All the following statements concerning the halogen family are CORRECT EXCEPT

a. bromine is obtained from bromide ion by using chlorine as the oxidizing agent.
b. iodine is the weakest oxidizing agent.
c. the halide ions are all reducing agents with the fluoride ion the weakest.
d. the halide ions are more reactive than the halogens.
e. all of the halogens can be obtained by electrolysis of either an aqueous solution or of the molten halide salts.

16.60 Which of the following statements concerning the halogen family (fluorine, chlorine, bromine and iodine) is FALSE?

a. As NaX salts, F^- is the poorest reducing agent.
b. All are oxidizing agents.
c. Fluorine is the most electronegative element.
d. The commercial production of HX is by the reaction of X_2 with water.
e. Bromine can be obtained from the ocean.

16.61

Group VII A	Boiling Point °C
F_2	-188
Cl_2	- 34.7
Br_2	58
I_2	183

The trend in boiling points is best explained by

a. an increase in van der Waals forces with increasing atomic number.
b. an increase in X-X bond energy with increasing atomic number.
c. an increase in the atomic mass with increasing atomic number.
d. a decrease in the ionization energy with increasing atomic number.
e. a decrease in the electronegativity with increasing atomic number.

40% 16.59 (d) The halogens (the elements) are much more reactive. The elements are not found in nature in the free state, but only in combined form. How long has NaCl been present in the Great Salt Lake?

16.60 (d) Cl_2, Br_2 and I_2 react slightly with water to give a mixture of HX and HOX. So this isn't a method to produce HX when there are so many other methods available that yield mostly product.

16.61 (a) Boiling points depend upon the strength of the <u>inter</u>molecular forces (forces between molecules not within the molecules).

16.62 Which of the following free halogens CANNOT be prepared by the electrolysis of a water solution of its salt?

a. F_2 b. Cl_2 c. Br_2 d. I_2 e. At_2

16.63 Which of the following is the most powerful reducing agent?

a. O^{2-} b. F^- c. Cl^- d. Br^- e. I^-

16.64 An atom of chlorine has the electronic configuration $1s^2\ 2s^2\ 2p^6\ 3s^2\ 3p^5$. The number of orbitals occupied by at least one electron is

a. 1 b. 5 c. 9 d. 13 e. 17

16.65 All of the following species would be expected to function as both oxidizing agents and reducing agents EXCEPT

a. ClO^- b. Cl_2 c. ClO_2^- d. Cl^- e. NO_2

16.66 Which of the following is obtained commercially by electrolytic <u>oxidation</u>?

a. Bromine b. Chlorine c. Magnesium d. KI e. Ethyl Alcohol

16.67 All of the equations below are examples of self oxidation and reduction reactions EXCEPT

a. $3NO_2 + H_2O \rightarrow 2HNO_3 + NO$

b. $3OH^- + P_4 + 3H_2O \rightarrow 3H_2PO_2^- + PH_3$

c. $Br_2 + H_2O \rightarrow HOBr + H^+ + Br^-$

d. $Br_2 + 2OH^- \rightarrow OBr^- + Br^- + H_2O$

e. All of the equations are self oxidation and reduction.

16.68 All of the following would be expected to undergo disproportionation EXCEPT

a. Br_2 b. BrO^- c. BrO_2^- d. Br^- e. BrO_3^-

16.62 (a) Fluorine must be prepared from the molten salt. In the electrolysis of NaF(aq), water will be oxidized in preference to the oxidation of fluoride ion.

16.63 (e) Which ion can lose an electron most readily? A large ion which has a low electronegativity.

16.64 (c) 1s (1) + 2s (1) + 2p (3) + 3s (1) + 3p (3)

1 + 1 + 3 + 1 + 3 = 9

16.65 (d) Cl^- can only function as a reducing agent since it is in its lowest oxidation state. All the other species are in intermediate oxidation states.

16.66 (b) Chlorine, $2\ Cl^- \rightarrow Cl_2 + 2\ e$, is obtained by oxidation and it in turn is used to obtain bromine. Magnesium is obtained by electrolytic reduction.

48% 16.67 (e) In disproportionation reactions, one reactant is both oxidized and reduced.

16.68 (d) Br^- can only be oxidized, it cannot be reduced further. All the other bromine species listed have bromine in an intermediate oxidation state.

16.69 Which one of the following has the smallest atomic radius?

a. Li b. Cs c. F d. Na e. Cl

16.70 Which of the following nonmetal oxides is paired with the acid which will result from the addition of water to the nonmetal oxide?

a. NO_2 and HNO_2 b. SO_2 and H_2SO_4 c. P_4O_{10} and H_3PO_3

d. Cl_2O_7 and $HClO_4$ e. NO and HNO_2

16.71 Of the following acids, the one with the strongest conjugate base is

a. HCl b. HClO c. $HClO_2$ d. $HClO_3$ e. $HClO_4$

16.72 Aqueous solutions of one of the following salts will give the most basic solution

a. NaOCl b. NaCl c. $NaClO_2$ d. $NaClO_3$ e. $NaClO_4$

16.73 Of the following noble gases, which reacts most readily with fluorine?

a. He b. Ne c. Ar d. Kr e. Xe

16.74 All of the following halides would be expected to hydrolyze rapidly EXCEPT

a. CF_4 b. SiF_4 c. SF_4 d. PF_3 e. PF_5

16.75 All of the following are ways that hydrogen chloride can be produced EXCEPT

a. $SiCl_4 + 2H_2O \rightarrow SiO_2 + 4HCl$

b. $CaCl_2 + H_2SO_4 \rightarrow CaSO_4 + 2HCl$

c. $H_2 + Cl_2 \xrightarrow{uv} 2HCl$

d. $C_2H_6 + Cl_2 \rightarrow C_2H_5Cl + HCl$

e. $KCl + H_2O \rightarrow KOH + HCl$

16.69 (c) The size of the atoms decreases in going from left to right in a period. Size increases down a family.

16.70 (d) Look for reaction in which there is no change in oxidation number.

$$Cl_2O_7 + H_2O \rightarrow 2\ HClO_4$$

Chlorine has a 7+ oxidation number in both the reactant and in the product.

16.71 (b) Which acid is the weakest acid? The weakest acid has the strongest conjugate base.

16.72 (a) The salt which has come from the weakest acid will have the strongest conjugate base. The strongest conjugate base will react more with water,

$$OCl^-(aq) + H_2O \rightarrow HOCl(aq) + OH^-(aq)$$

16.73 (e) The noble gas (the largest one) which has the lowest ionization energy.

68% 16.74 (a) Carbon in CF_4 cannot readily form a five bonded intermediate, whereas all the other central atoms can form more bonds. For example, all the following species are known: SiF_6^{2-}, SF_6, PF_5, PF_6^-.

30% 16.75 (e) KCl is a salt of a strong base and a strong acid. The chloride ion does not react with water because it is a very weak conjugate base.

16.76 In selecting a strong oxidizing agent for oxidizing Sn^{2+} to Sn^{4+}, one should not consider

a. Hot 18 M H_2SO_4 b. Hot 12 M HCl c. 12 M HNO_3 d. $Cl_2(g)$

e. $Br_2(\ell)$

16.77 The correct Lewis structure for chlorous acid is

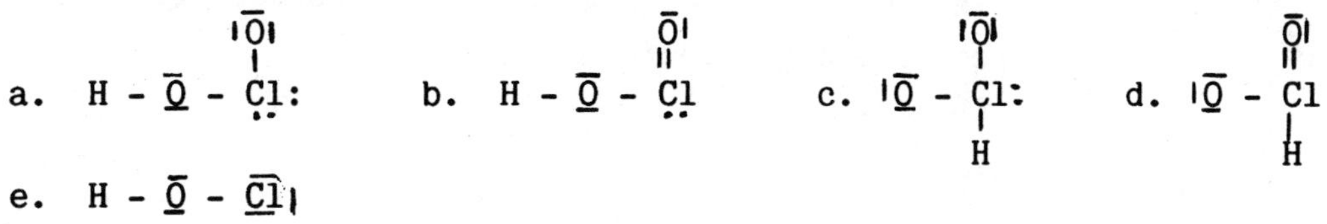

16.78 All the following are characteristic of nonmetal halides EXCEPT

a. polar covalent bonds.

b. polar molecules are formed with family VA.

c. the central nonmetal atom has a lower oxidation state than the halogen.

d. hydrolyze rapidly.

e. gases, liquids, or low melting solids.

16.79 Chloride ions would be expected to be oxidized to chlorine in acid solution by all of the following chemicals EXCEPT

a. MnO_2 b. $KMnO_4$ c. $K_2Cr_2O_7$ d. $KClO_3$ e. K_2O

16.76 (b) The chloride ion is not an oxidizing agent and H^+ is a very weak oxidizing agent.

36% 16.77 (a) Chlorous acid is $HClO_2$. The hydrogen is bonded to an oxygen atom. Structure (b) is not correct because there are not enough electrons.

64% 16.78 (c) The central atom has a positive oxidation state whereas the halogen is assigned an oxidation state of minus one.

60% 16.79 (e) K_2O is a basic anhydride. K^+ is an extremely weak oxidizing agent.

17

TRANSITION ELEMENTS

17.1 Transition metals can be distinguished from main group metals by the fact that

a. main group metals only have +1 or +2 oxidation states.

b. main group metals have higher relative atomic masses than transition metals.

c. transition metals have higher relative atomic masses than the main group metals.

d. only the main group metals can form complex ions.

e. transition metals have a greater tendency to form colored compounds than main group metals.

17.2 The property which is common to many representative nonmetals and many transition elements is that they have

a. high electronegativities.

b. color in many compounds.

c. many oxidation states.

d. many paramagnetic compounds.

e. many compounds which observe the octet rule.

17.3 All the properties listed below are characteristic of many of the first transition element compounds EXCEPT

a. all of the elements have a plus two oxidation state.

b. most of the complexes are colored.

c. a few of the complexes are diamagnetic.

d. most of the elements exhibit multiple oxidation states.

e. the maximum oxidation state for the first few elements usually corresponds to the group number.

17.4 All the properties listed below are characteristic of many of the transition element compounds EXCEPT

a. most of them are paramagnetic.

b. most of them are colored.

c. most of the elements upon ionizing lose the d electrons first.

d. most of the metals exhibit multiple oxidation states.

e. most of the elements form many complexes.

84% 17.1 (e) Transition metal compounds which have d^1 to d^9 electrons are colored. Most compounds or ions that have noble gas configurations are colorless.

17.2 (c) (a) Transition elements have low electronegativities.
(b) Most representative element compounds are not colored.
(d) Most representative element compounds are diamagnetic.
(e) Most transition element compounds violate the octet rule.

68% 17.3 (a) Scandium only has a plus three oxidation state.

60% 17.4 (c) The s electrons are lost first and then the d electrons. The electronic configuration of Ti^{2+} is $4s^0\ 3d^2$ not $4s^2\ 3d^0$.

17.5 The following elements have the listed oxidation states EXCEPT

a. Sc^{2+} b. Sc^{3+} c. Ti^{2+} d. Ti^{3+} e. V^{2+}

17.6 The highest oxidation state for chromium is

a. +2 b. +3 c. +5 d. +6 e. +7

17.7 The maximum oxidation state for vanadium, $_{23}V$, is

a. 2 b. 3 c. 4 d. 5 e. 6

17.8 All of the following would be expected to be strong oxidizing agents EXCEPT

a. $KMnO_4$ b. $H_2Cr_2O_7$ c. HNO_3 d. KO_2 e. Sc_2O_3

17.9 All of the oxides given below would be expected to react with oxygen EXCEPT

a. NO b. N_2O c. VO d. V_2O_3 e. TiO_2

17.10 All of the following would be expected to function as reducing agents EXCEPT

a. NH_3 b. V^{3+} c. Zn^{2+} d. P_4 e. H_2

17.11 Considering only electronic structures, all of the following should be colorless EXCEPT

a. $CoCl_2$ b. $ZnCl_2$ c. Na_2SO_4 d. $CaCl_2$ e. $AlCl_3$

17.12 Which one of the following compounds would be expected to be colorless?

a. ZnS b. CuS c. NiS d. FeS e. CoS

17.13 From a consideration of the electronic configuration, all of the following compounds would be expected to be paramagnetic EXCEPT

a. $NiCl_2$ b. $CoSO_4$ c. $VOCl_3$ d. TiF_3 e. Mn_2O_3

17.5 (a) Scandium is in Family III and it only has the 3+ oxidation number.

80% 17.6 (d) This corresponds to Group Number VI to which chromium belongs.

60% 17.7 (d) Vanadium belongs to Group V, so this is its maximum oxidation state.

56% 17.8 (e) Transition elements that are in relatively high oxidation states are many times strong oxidizing agents.

72% 17.9 (e) Titanium is already in its highest oxidation state.

64% 17.10 (c) Zn^{2+} is in Family II and is in its highest oxidation state so it cannot give up any more electrons.

84% 17.11 (a) $CoCl_2$ has some d electrons (any element from d^1 to d^9) so it is colored.

80% 17.12 (a) Zn^{2+} in ZnS is a d^{10} ion. Therefore, it does not have any unpaired electrons.

17.13 (c) The transition elements which have d^1 to d^9 electrons can be paramagnetic. $VOCl_3$ is a d^0 case and it is diamagnetic.

17.14 The electron configuration of Mn^{2+} is

a. [Ar] $4s^2 3d^5$ b. [Ar] $4s^2 3d^3$ c. [Ar] $4s^2 4d^3$

d. [Ar] $4s^1 3d^4$ e. [Ar] $3d^5$

17.15 The Cr^{3+} ion would be expected to have ___ unpaired electrons.

a. 0 b. 1 c. 2 d. 3 e. 4

17.16 Considering electronic structures, one would predict that all of the following compounds would be colorless EXCEPT

a. $ZnCl_2$ b. $MgSO_4$ c. NH_3 d. $TiCl_4$ e. ClO_2

17.17 The oxidation state of Re in $Ca(ReO_4)_2$ is

a. 3 b. 4 c. 5 d. 6 e. 7

17.18 The cobalt atom would be expected to have ___ unpaired electrons.

a. 0 b. 1 c. 2 d. 3 e. 4

17.19 An ion that has a similar valence electron configuration to that of Zn^{2+} is

a. Cu^{2+} b. Ga^+ c. Ca^{2+} d. Pb^{2+} e. Ge^{2+}

17.20 From a consideration of electronic structures, all the following compounds would be expected to be colored EXCEPT

a. ZnO b. TiO c. Ti_2O_3 d. MnO_2 e. Mn_2O_3

17.21 Reduction of MnO_4^- in basic solution produces

a. Mn b. Mn^{2+} c. Mn^{3+} d. MnO_2 e. Mn_2O_3

17.14 (e) The s electrons are lost first.

60% 17.15 (d) Cr^{3+} $\underline{\ \ }$ $\underline{\uparrow}$ $\underline{\uparrow}$ $\underline{\uparrow}$ $\underline{\ \ }$ $\underline{\ \ }$

4s 3d

Remember the 4s electrons are removed first.

17.16 (e) Look for a transition element that has d^1 to d^9 electrons. Since none of the transition elements qualify, look for a compound consisting of representative elements which has an odd number of electrons.

17.17 (e) Since calcium is in Family II, it has a 2+ charge. Therefore, ReO_4 must be minus one ion.

$\therefore\ y + 4(-2) = -1$

17.18 (d) Co $4s^2\ 3d^7$; $\underline{\uparrow\downarrow}$ $\underline{\uparrow\downarrow}$ $\underline{\uparrow\downarrow}$ $\underline{\uparrow}$ $\underline{\uparrow}$ $\underline{\uparrow}$

4s 3d

44% 17.19 (c) Ca^{2+} is $4\ s^o$

Zn^{2+} is $4\ s^o\ 3d^{10}$

After the d orbitals are filled, the d electrons are no longer valence electrons.

17.20 (a) A metal ion has from 1 to 9 d electrons will be colored, but Zn^{2+} is $4s^o\ 3d^{10}$ and it is colorless.

17.21 (d)

17.22 Which reaction is the most important for producing iron atoms in a blast furnace?

a. $2Fe_2O_3 \rightarrow 4Fe + 3O_2$

b. $Fe_2O_3 + 3C \rightarrow 2Fe + 3CO$

c. $2Fe_2O_3 + 3C \rightarrow 2Fe + 3CO_2$

d. $Fe_2O_3 + 3CO \rightarrow 2Fe + 3CO_2$

e. $Fe_2O_3 + 2Al \rightarrow 2Fe + Al_2O_3$

17.23 Ag^+ forms many complexes with a coordination number of two. Some of these are $Ag(NH_3)_2^+$, $Ag(CN)_2^-$, $Ag(S_2O_3)_2^{3-}$. Which of the following statements is FALSE?

a. In these complexes, Ag^+ is a Lewis acid.

b. These complexes are all linear.

c. These complexes are paramagnetic.

d. The hybridization of Ag^+ is sp.

e. The Ag^+ complexes can function as oxidizing agents.

17.24 Ag^+ forms many complexes with a coordination number of two. Some of these are $Ag(NH_3)_2^+$, $Ag(CN)_2^-$, $Ag(S_2O_3)_2^{3-}$. Which of the following statements is true?

a. These complexes are all linear.

b. In these complexes, Ag^+ is a Lewis base.

c. The hybridization of Ag^+ is sp^2.

d. The hybridization of Ag^+ is dsp.

e. The Ag^+ complexes are good reducing agents.

52% 17.22 (d) Carbon monoxide is the effective reducing agent because the gas molecules can diffuse through the mineral and interact with the iron ions. The carbon in the furnace is converted to carbon monoxide. Solid-solid reactions are difficult to carry out.

60% 17.23 (c) The silver complexes are diamagnetic because the silver ion has the electronic configuration $5s^0\ 4d^{10}$.

17.24 (a) Compare question 17.24 with question 17.23. There are two electron pairs about the central silver ion and no lone pairs of electron so it is linear.

For the next four questions, select the best one of the reagents listed to separate (precipitate one and not the other or dissolve one solid and not the other) the ions or solids given. A reagent may be selected more than one time.

17.25 $Ag^{+}(aq)$ from $Cu^{2+}(aq)$

17.26 $Cu^{2+}(aq)$ from $Zn^{2+}(aq)$

17.27 $SnS_2(s)$ from $CuS(s)$

17.28 $AgCl(s)$ from $Hg_2Cl_2(s)$

Responses for questions 17.25 - 17.28.

a. hot water

b. dilute HCl

c. dilute NH_3 water

d. 0.3 M HCl saturated with H_2S

e. 6 M NaOH and 2 M NaHS

17.25 (b) All common chlorides are soluble except AgCl, $PbCl_2$ and Hg_2Cl_2.

$$Ag^+ + Cl^- \rightarrow AgCl(s)$$

17.26 (d) CuS is much more insoluble than ZnS. Zinc sulfide can be dissolved by adding dilute hydrochloric acid to it. In 0.3 M HCl saturated with H_2S, the S^{2-} concentration is not high enough to precipitate ZnS.

17.27 (e) $HS^- + OH^- \rightarrow S^{2-} + H_2O$

In the presence of excess sulfide ion SnS_2 forms a complex sulfide and dissolves whereas CuS does not react.

$$SnS_2 + S^{2-} \rightarrow SnS_3^{2-}$$

17.28 (c) We will get a "messy" separation with NH_3.

$$AgCl(s) + 2\ NH_3(aq) \rightarrow Ag(NH_3)_2^+ + Cl^-$$

$$Hg_2Cl_2(s) + NH_3 \rightarrow Hg + HgNH_2Cl(s) + H^+ + Cl^-$$

18

ORGANIC CHEMISTRY

18.1 The existence of thousands of different hydrocarbons is because

a. carbon exists in allotropic forms.

b. carbon can form four bonds.

c. carbon forms strong bonds with both itself and hydrogen.

d. carbon forms strong bonds with itself and weak bonds with hydrogen.

e. carbon forms weak bonds with itself and strong bonds with hydrogen.

18.2 The existence of millions of different organic compounds is best accounted for by the reason(s)

a. carbon forms strong bonds with hydrogen and other nonmetallic elements.

b. carbon forms four bonds and has no lone pair of electrons or empty valence orbitals.

c. carbon forms strong single bonds and stronger multiple bonds with itself.

d. carbon forms strong single bonds as well as stronger double bonds with oxygen.

e. All of the above.

18.3 The chemical structure

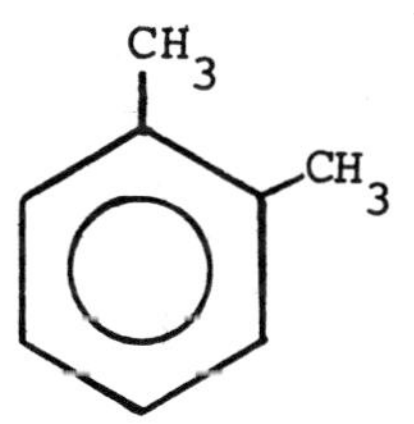

represents a molecule of molecular formula:

a. C_8H_{10} b. C_6H_6 c. C_6H_8 d. C_8H_{12} e. C_8H_6

72% 18.1 (c) Carbon is a relatively small atom that has an electronegativity value similar to hydrogen. The C-C bond energy is roughly of the same magnitude as the energies of the bonds that carbon forms to other elements. See question 18.2.

68% 18.2 (e)

60% 18.3 (a) Each intersection in the hexagon represents a carbon that has 4 bonds. At 4 of the intersections, the carbon is bonded to hydrogen.

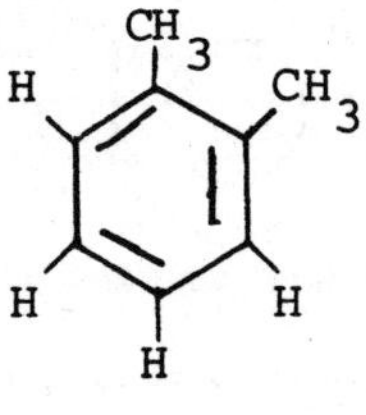

18.4 The chemical structure

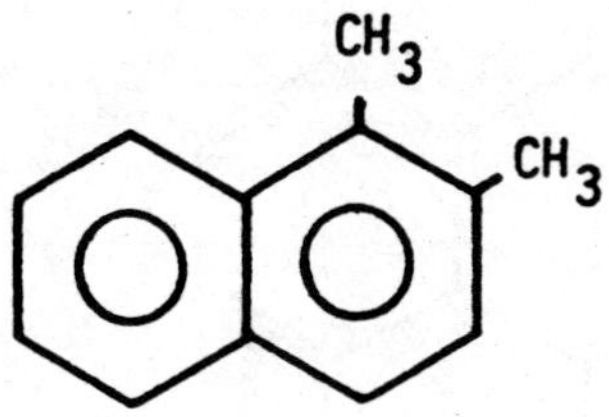

represents a molecule of molecular formula

a. $C_{10}H_{14}$ b. $C_{10}H_{16}$ c. $C_{12}H_{12}$ d. $C_{12}H_{14}$ e. $C_{12}H_{16}$

18.5 Four of the following compounds are structural isomers. Which one is not one of these isomers?

a. 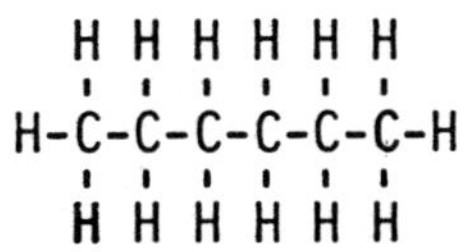

b. 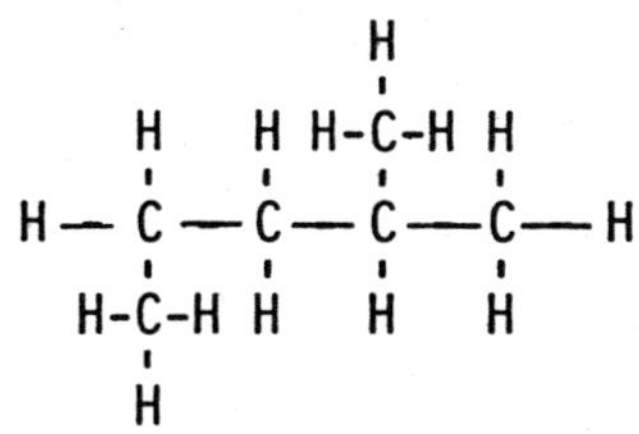

c. 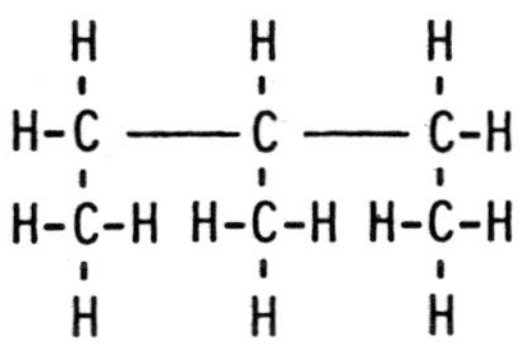

d. 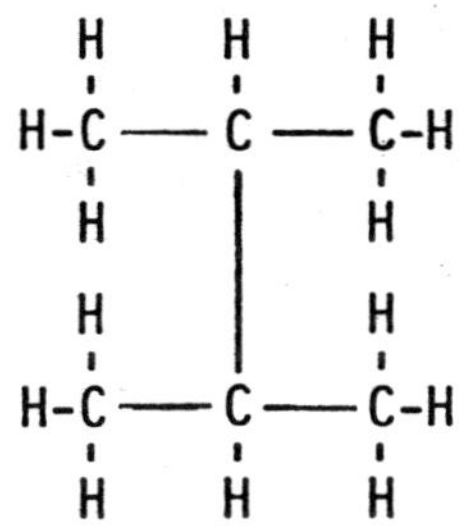

e. 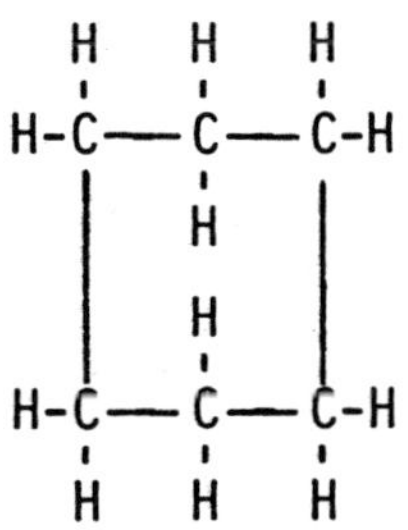

80% 18.4 (c)

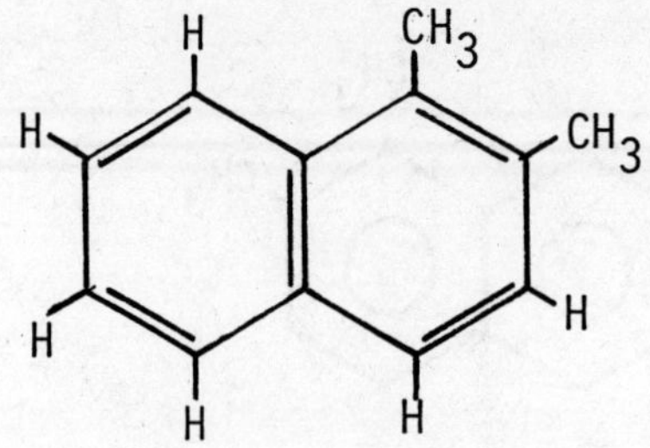

Each carbon has four bonds.

92% 18.5 (e) Noncyclic saturated hydrocarbons have the general formula C_nH_{2n+2}. Cyclic saturated hydrocarbons have the general formula C_nH_{2n} so compound (e) cannot be an isomer.

18.6 How many different isomers of C_6H_{14} are shown below?

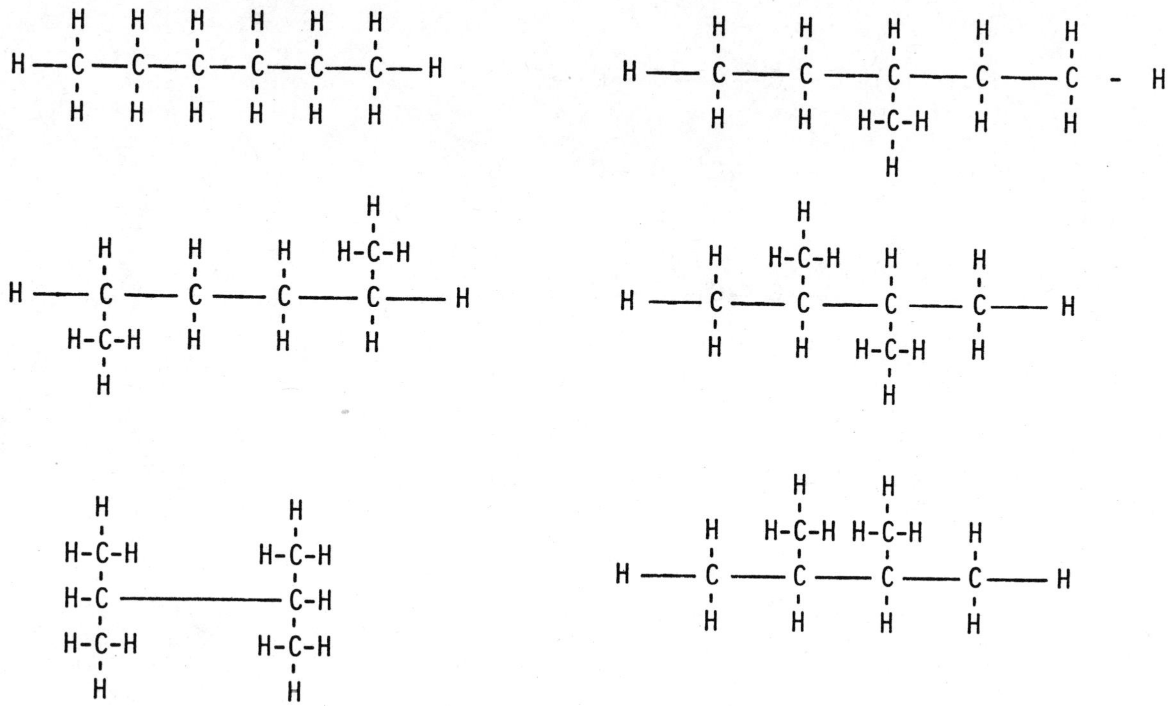

a. 1 b. 2 c. 3 d. 4 e. 5

18.7 The total number of isomeric alkanes of the molecular formula C_6H_{14} is

a. 1 b. 3 c. 5 d. 6 e. 7

18.8 The formulas of a structural (functional) isomer of 2-butyne is (are)

1. $CH_3CH{\equiv}CHCH_3$ 2. $CH_2 = C = CHCH_3$ 3. $CH_2 = CHCH = CH_2$

4. 

a. 1 only b. 2 only c. 2 and 3 only d. 1, 2 and 3 only
e. 2, 3 and 4 only

18.9 Which of the following pairs of compounds are isomers?

a. $CH_3CH_2CH_3$ and CH_3CHCH_3 with CH_3 on the middle carbon b. $CH_2 = CH_2$ and $CH{\equiv}CH$

c. $CH_3\overset{O}{\overset{\|}{C}}OH$ and $H\overset{O}{\overset{\|}{C}}OCH_3$ d. $CH_3CHOHCH_3$ and $CH_3\underset{O}{\underset{\|}{C}}CH_3$

e. CH_3OCH_3 and CH_2—CH_2 (bridged by O)

46% 18.6 (c)

18.7 (c)

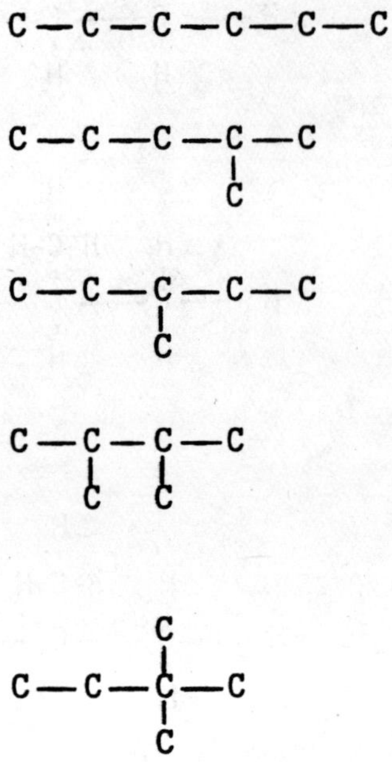

30% 18.8 (e) Compound 1. is nonsense as there are too many hydrogen atoms and/or bonds around the central carbon atoms.

2-butyne is $CH_3C{\equiv}CCH_3$ (C_4H_6)

All the structural isomers have the formula C_4H_6 including compound 4.

Compound 4 is

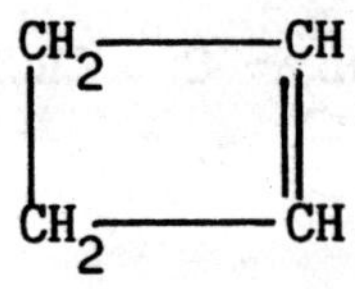

92% 18.9 (c) In (c) we have functional isomers as there are the same number of atoms ($C_2H_4O_2$) but different functional groups.

18.10 Geometric isomerism is usually found in

a. alkanes b. alkenes c. alkynes d. alcohols e. aldehydes

18.11 The isomeric hexane that has the highest boiling point is

a. $CH_3CH_2CH_2CH_2CH_2CH_3$ b. $CH_3CH_2CH_2\underset{CH_3}{CH}CH_3$ c. $CH_3CH_2\underset{CH_3}{CH}CH_2CH_3$

d. $CH_3\underset{CH_3}{CH}\,\underset{CH_3}{CH}\,CH_3$ e. $CH_3CH_2\overset{CH_3}{\underset{CH_3}{C}}CH_3$

18.12 How many different isomeric alcohols have the molecular formula C_4H_9OH? (Be sure to count each pair of optical isomers as two isomers.)

a. 1 b. 3 c. 4 d. 5 e. 6

18.13 Which one of the following species would abe expected to show optical activity?

a. $HO\overset{OCH_3}{\underset{CH_3}{C}}CO_2H$ b. $H\underset{NH_2}{C}H\text{-}CO_2H$ c. $CH_3\overset{H}{C}HCO_2H$ d. $H\overset{Cl}{\underset{Cl}{C}}Br$

e. $H\overset{\overset{O}{\|}}{C}OCH_3$

18.14 The correct systematic name of isoamyl alcohol, $(CH_3)_2CHCH_2CH_2OH$ is

a. isoamyl alcohol b. isopentyl alcohol c. 3-methyl-1-butanol

d. 3,3-dimethyl-1-propanol e. 2-isopropyl-1-ethanol

50% 18.10 (b) In order to have cis and trans isomers there needs to be double bond in order to prevent rotation of the molecule about the bonding axis.

64% 18.11 (a) The completely linear alkanes are able to pack together more closely so the intermolecular bonding forces are stronger. Branching of the hydrocarbon (side chains) increases the distances between molecules and thus decreases the intermolecular forces.

44% 18.12 (d) C - C - C - C - OH

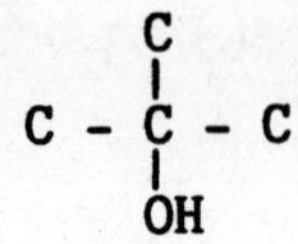

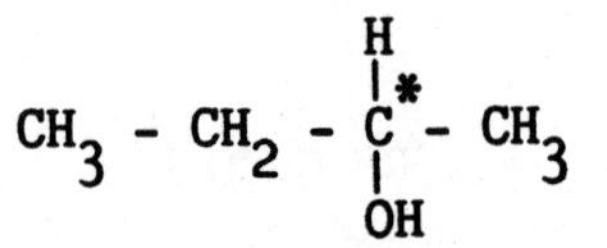

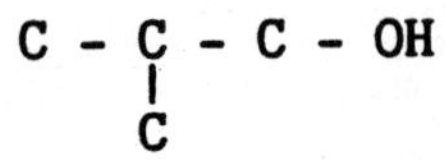

d-ℓ pair

60% 18.13 (a) Four different groups must be bonded to carbon in order for it to be asymmetric.

18.14 (c) The longest continuous chain is four carbons, so the compound should be named as a substituted butane.

18.15 The correct name for 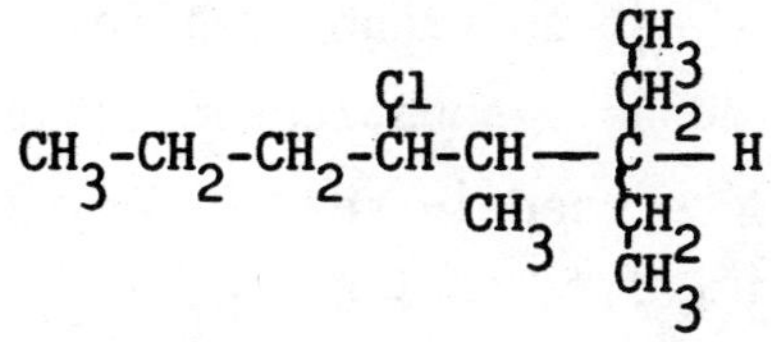 is

a. 3-chloro-1, 1-diethyl-2-methylhexane

b. 4-chloro-6-ethyl-5-methyloctane

c. 5-chloro-3-ethyl-4-methyloctane

d. 4-chloro-1, 1-diethyl-5-methylhexane

e. None of these.

18.16 The correct structure of meta dichlorobenzene is

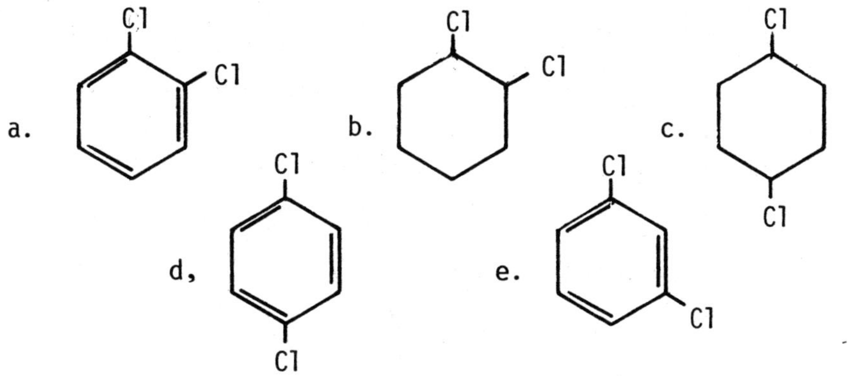

The species below are to be used for questions 18.17 - 18.21.

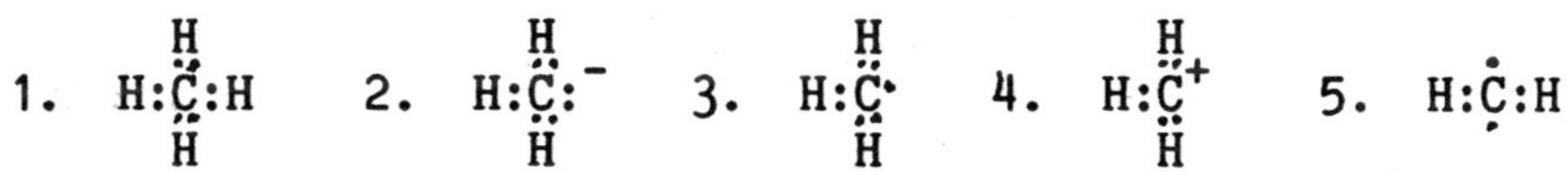

18.17 Which would be expected to be the best Lewis acid?

a. 1 b. 2 c. 3 d. 4 e. 5

18.18 Which would be expected to be the best Lewis base?

a. 1 b. 2 c. 3 d. 4 e. 5

18.19 Which would be expected to be pyramidal in shape?

a. 1 b. 2 c. 3 d. 4 e. 5

18.20 Which species represents a methyl free radical?

a. 1 b. 2 c. 3 d. 4 e. 5

56% 18.15 (c) The longest continuous chain is eight carbons, so it should be named as a substituted octane. The smallest numbers should be used in the numbering of the side groups.
$(5 + 3 + 4) < (4 + 6 + 5)$

60% 18.16 (e) Meta dichlorobenzene (1,3 disubstituted)
Para dichlorobenzene (1,4 disubstituted)
Ortho dichlorobenzene (1,2 disubstituted)

70% 18.17 (d) A Lewis acid is an electron pair acceptor. The acid has available an empty valence orbital.

76% 18.18 (b) A Lewis base is an electron pair donor. The base has a lone pair of valence electrons available for donation.

40% 18.19 (b) Three atoms bonded to a central atom with a lone pair of electrons provides four stereoactive sets of electrons.

18.20 (c)

18.21 Which would be expected to have the carbon hybridized sp^2?

a. 1 b. 2 c. 3 d. 4 e. 5

18.22 All of the carbon atoms which are designated with an asterick are sp^2 hybridized EXCEPT

a. $CH_3\underset{*}{\overset{\overset{O}{\|}}{C}}-H$ b. $CH_3\underset{*}{\overset{\overset{O}{\|}}{C}}-OH$ c.  * d. $\underset{*}{C_2}H_2$ e. $Cl_2\underset{*}{C}O$

18.23 The functional group $-\overset{\overset{O}{\|}}{C}-H$ is characteristic of organic

a. ketones b. acids c. aldehydes d. esters e. alcohols

18.24 The functional group $-\overset{\overset{O}{\|}}{C}-OH$ is characteristic of organic

a. ketones b. acids c. aldehydes d. esters e. alcohols

18.25 The functional group $-\overset{\overset{O}{\|}}{C}-O-R$ is characteristic of organic

a. ketones b. acids c. aldehydes d. esters e. alcohols

18.26 The functional group $R-\overset{\overset{O}{\|}}{C}-R$ is characteristic of organic

a. ketones b. acids c. aldehydes d. esters e. alcohols

18.27 The functional group of an alkyne consists of

a. 1 σ bond and 2 π bonds.

b. 1 σ bond and 1 π bond.

c. 2 σ bonds and 1 π bond.

d. 3 σ bonds.

e. 3 π bonds.

18.28 Which of the following organic compounds would have the highest boiling point?

a. CH_3CH_2OH b. CH_3CH_2Cl c. CH_3OCH_3 d. CH_3CHO e. $CH_3CH_2CH_3$

36% 18.21 (d)

18.22 (d) The carbon atoms in C_2H_2 are sp hybridized.

18.23 (c)

18.24 (b)

18.25 (d)

18.26 (a)

80% 18.27 (a) $C \equiv C$
Alkyne functional group has three bonds.

18.28 (a) All of the molecules are about the same size. A question that should be asked is whether there is any hydrogen bonding. In acetaldehyde (ethanal), CH_3CHO, the hydrogen is bonded to carbon and not oxygen so there is no hydrogen bonding.

18.29 Oxidation of an aldehyde may yield a(an)

a. ketone b. alkene c. alcohol d. acid e. ether

18.30 The reagents for the preparation of an ester could be

a. an organic acid.

b. an alcohol and a ketone.

c. an ether and an alcohol.

d. an organic acid and an alcohol.

e. an alkene and an ether.

18.31 All of the following statements are TRUE EXCEPT

a. Excess propane reacts with chlorine in the presence of ultraviolet radiation to produce 1-chloropropane and 2-chloropropane.

b. Propane can be burned in oxygen.

c. Propene reacts with chlorine in the dark.

d. Methane reacts with chlorine in the presence of ultraviolet radiation to yield an isomeric mixture.

e. The products of the oxidation of methane are carbon dioxide and water.

18.32 All of the following describe benzene EXCEPT

a. an aromatic hydrocarbon.

b. three double bonds.

c. resonance hybrid structures.

d. undergoes substitution reactions.

e. can have ortho-, meta-, and para-disubstituted products.

18.33 The addition of HCl to $(CH_3)_2-C{=}CHCH_3$ yields

a. $(CH_3)_2CHCH_2CH_3$ b. $(CH_3)_2-\underset{H}{C}-\overset{H}{\underset{Cl}{C}}-CH_3$ c. $(CH_3)_2-\underset{Cl}{C}-\overset{H}{\underset{H}{C}}-CH_3$

d. $(CH_3)_2-\underset{H}{C}-\overset{H}{\underset{H}{C}}-CH_2Cl$ e. $CH_3-\overset{H}{\underset{CH_2Cl}{C}}-CH_2-CH_3$

45% 18.29 (d) For example

$$CH_3CHO \xrightarrow{[O]} CH_3COOH$$

70% 18.30 (d) For example

$$CH_3COOH + C_2H_5OH \xrightarrow{[H^+]} CH_3COOC_2H_5 + H_2O$$

35% 18.31 (d) Methane reacts with chlorine to give the different compounds, CH_3Cl, CH_2Cl_2, $CHCl_3$ and CCl_4, but no isomers.

60% 18.32 (b) There is delocalization of the bonding electrons, so effectively the carbon-carbon bonds are 1.5 bonds.

70% 18.33 (c) Addition of an unsymmetrical reagent to an unsymmetrical alkene follows Markovnikov's rule. The positive part of the addend becomes attached to the double-bonded carbon atom which has the greatest number of hydrogen atoms.

18.34 Hydrolysis of an ester would yield

a. an organic acid only.

b. a soap.

c. an organic acid and an alcohol.

d. an alcohol and sodium hydroxide.

e. saponification.

18.35 In which of the following systems would reaction occur by substitution?

a. $CH_2 = CH_2 + H_2 \xrightarrow{\text{catalyst}}$

b.  $+ Br_2 \xrightarrow{\text{catalyst}}$

c.  $+ Br_2 \longrightarrow$

d. $CH_3\overset{O}{\overset{\|}{C}}OH + NaOH \longrightarrow$

e.  $+ HBr \longrightarrow$

18.36 Ethanol can be oxidized stepwise to two stable intermediate products. One of these intermediate products is

a. CH_3CH_2OOH b. CO_2 c. CH_3CHO d. CH_3CH_2OH e. CH_3OCH_3

18.37 All of the following compounds would be expected to polymerize EXCEPT

a. H_2CCHCl b. F_2CCF_2 c. H_3CCH_3 d. $H_2NCH_2CH_2COOH$

e. $H_2N(CH_2)_4NH_2 + HOOC(CH_2)_4COOH$

18.38 The main elements of proteins are

a. carbon and hydrogen

b. carbon, hydrogen and oxygen

c. carbon, oxygen, nitrogen and hydrogen

d. carbon and oxygen

e. carbon, oxygen, hydrogen and sulfur

75% 18.34 (c) Hydrolysis or saponification of an ester gives back the reagents that produce the ester. See question 18.30.

60% 18.35 (b) Most reactions involving an aromatic ring are electrophilic substitution reactions.

$$C_6H_6 + Br_2 \xrightarrow{FeBr_3} C_6H_5Br + HBr$$

50% 18.36 (c) $C_2H_5OH \xrightarrow{[O]} CH_3CHO \xrightarrow{[O]} CH_3COOH \xrightarrow{[O]} CO_2$

60% 18.37 (c) For polymerization to take place, each monomer must have two reactive sites. Ethane lacks even a functional group let alone two reactive sites.

18.38 (c)

Cations

NH_4^+ ammonium H_3O^+ Hydronium

Anions

OH^-	Hydroxide
CN^-	Cyanide
$C_2H_3O_2$	Acetate
CO_3^{2-}	Carbonate
HCO_3^-	Hydrogen Carbonate
PO_4^{3-}	Phosphate
MnO_4^-	Permanganate
$Cr_2O_7^-$	Dichromate
ClO_4^-	Perchlorate
ClO_3^-	Chlorate
CrO_4^{2-}	Chromate
NO_2^-	Nitrite
NO_3^-	Nitrate
SO_3^{2-}	Sulfite
SO_4^{2-}	Sulfate
HSO_4^-	Hydrogen sulfate
HPO_4^{2-}	Monohydrogen Phosphate
$H_2PO_4^-$	Dihydrogen Phosphate

Common Acids and Bases

Acids

HCl	Hydrochloric Acid	(strong)
HNO_3	Nitric Acid	(strong)
$HClO_4$	Perchloric Acid	(Strong)
H_2SO_4	Sulfuric Acid	(Strong)
H_3PO_4	Phosphoric Acid	(Weak)
$HC_2H_3O_2$	Acetic Acid	(Weak)
H_2CO_3	Carbonic Acid	(Weak)

Bases

$NaOH$	Sodium Hydroxide
KOH	Potassium Hydroxide
$Ca(OH)_2$	Calcium Hydroxide
NH_3	Ammonia
$LiOH$	Lithium Hydroxide
$Ba(OH)_2$	Barium Hydroxide

Soluble Compounds

Soluble Compounds	Exceptions
Almost all sodium (Na^+), potassium (K^+), and ammonium (NH_4^+) salts	
All chlorides (Cl^-), bromides (Br^-), iodides (I^-) →	Ag^+, Hg_2^{2+}, Pb^{2+}
All flouride (F^-) →	Mg^{2+}, Ca^{2+}, Sr^{2+}, Ba^{2+}, Pb^{2+}
All nitrates (NO_3^-), chlorates (ClO_3^-), perchlorates (ClO_4^{2-}) and acetates ($C_2H_3O_2^-$) →	Acetates of Ag^+ and Hg_2^{2+} are only moderately so
All sulfates (SO_4^{2-}) →	Sr^{2+}; Ba^{2+}; Pb^{2+}; (Ca^{2+} and Ag^+ are moderately solu

Poorly Soluble Salts

Poorly Soluble Salts	Exceptions
All carbonates (CO_3^{2-}), phosphates (PO_4^{3-}), oxalates ($C_2O_4^{2-}$) and chromates (CrO_4^{2-}) →	Na^+, K^+, NH_4^+
All sulfides (S^{2-}) →	Alkali and Alkaline earth metal ions and N
All hydroxides (OH^-) and oxides (O^{2-}) →	Alkali metals (those of Ca, Sr, and Ba only moderately solubl